Implementing Typed Feature
Structure Grammars

Implementing Typed Feature Structure Grammars

Ann Copestake

CSLI PUBLICATIONS
Center for the Study of
Language and Information
Stanford, California

Copyright © 2002
CSLI Publications
Center for the Study of Language and Information
Leland Stanford Junior University
Printed in the United States
06 05 04 03 02   5 4 3 2 1

Library of Congress Cataloging-in-Publication Data

Copestake, Ann.
Implementing typed feature structure grammars /
Ann Copestake.
p.   cm. – (CSLI lecture notes ; no. 110)
Includes bibliographical references (p.   ) and index.

ISBN 1-57586-261-1 (alk. paper)
ISBN 1-57586-260-3 (pbk. : alk. paper)

1. Computational linguistics. I. Title. II. Series.
P98.C635   2001
410′.285—dc21   2001047608
CIP

∞ The acid-free paper used in this book meets the minimum requirements of
the American National Standard for Information Sciences—Permanence of
Paper for Printed Library Materials, ANSI Z39.48-1984.

CSLI was founded early in 1983 by researchers from Stanford University, SRI
International, and Xerox PARC to further research and development of integrated
theories of language, information, and computation. CSLI headquarters and CSLI
Publications are located on the campus of Stanford University.

CSLI Publications reports new developments in the study of language, information,
and computation. In addition to lecture notes, our publications include
monographs, working papers, revised dissertations, and conference proceedings.
Our aim is to make new results, ideas, and approaches available as quickly as
possible. Please visit our web site at
http://cslipublications.stanford.edu/
for comments on this and other titles, as well as for changes and corrections by the
author and publisher.

Contents

Preface

This book is an an introduction to practical grammar implementation based on the use of the Linguistic Knowledge Building (LKB) system. The LKB system was initially developed at the University of Cambridge Computer Laboratory as part of the ACQUILEX projects[1] The first version was implemented in 1991 primarily as a tool to allow the construction of typed feature structure based lexicons (both monolingual and bilingual). At this point, LKB stood for Lexical Knowledge Base — the LKB system was a tool for building LKBs. The current version of the LKB system (now renamed the 'Linguistic Knowledge Building' system, to reflect its use for all grammar components and not just lexicons) has been extensively updated, mostly at the Center for the Study of Language and Information (CSLI), Stanford University. This later work is part of the LinGO ('Linguistic Grammars Online') project and was partially supported by the National Science Foundation.[2]

The main developers of the LKB system besides myself are John Carroll, Rob Malouf and Stephan Oepen. The distributed system also incorporates code written by Bernie Jones, Dan Flickinger and John Bowler. Ted Briscoe was responsible for deciding to build the initial LKB system and he, Antonio Sanfilippo and Valeria de Paiva had a great deal of influence on its design. The first version also owed a considerable amount to Shieber's PATR-II. Work on the original LKB system formed part of my DPhil thesis at the University of Sussex, supervised by Gerald Gazdar, who guided several aspects of the approach to lexical representation (but was unfortunately unable to prevent some over-frequent revision cycles). Dan Flickinger was a primary instigator of the more recent work on the LKB, and played a major part in determining how

---

[1] ACQUILEX projects BRA-3030 and 7315 to the University of Cambridge under the EU-funded Esprit program.

[2] Grant number IRI-9612682 to Stanford University.

it developed, especially the attention given to efficiency and system robustness. Uli Callmeier uncovered numerous bugs and infelicities in the course of his own work on the PET system, and some of his algorithms have been incorporated into the LKB system.

The formal account of the initial LKB system was produced jointly with Valeria de Paiva. The approach to defaults now used in the LKB is based on work with Alex Lascarides, Ted Briscoe and Nicholas Asher. Both the formalization of typed feature structures and the way I have attempted to describe them informally owe a great deal to work by other authors, especially Stuart Shieber and Bob Carpenter. Besides the colleagues mentioned above, I am grateful to Bob Carpenter and Roger Evans for discussions about representation issues. Much of my understanding of the utilization of typed feature structures in HPSG comes from Ivan Sag. I have had many informative and sometimes lively discussions with developers and users of other formalisms and grammar development environments, which have helped me to clarify many ideas and, I hope, improve the way they are expressed here. But because this is intended as an introductory book, I hope these colleagues will forgive the fact that there is very little comparison with other systems and formalisms. The evolution of the LKB has of course been influenced by other systems: in particular, the syntax of the description language and one method of encoding irregular morphology are adapted from the PAGE system.

Invaluable comments, bug reports and suggestions have been provided by users far too numerous to thank individually, including colleagues on the ACQUILEX and LinGO projects, students and researchers at the University of Cambridge, Stanford University, Copenhagen Business School and Edinburgh University, and students on courses taught by Flickinger, Malouf, Oepen and myself at ESSLLI-98 and ESSLLI-00. Special thanks go to Matt Kodama, Ryan Ginstrom, Christopher Callison-Burch and Scott Guffey for their detailed comments on design and extensive testing. I am grateful to all LKB users for their tolerance in the face of the inevitable bugs and infelicities, in particular Fabre Lambeau who showed incredible patience in tracking down the mysterious NumLock problem.

The sample grammars described in this book are partly based on the ESSLLI course grammars which were written jointly with Flickinger, Malouf and Oepen. These in turn were partially based on the 'textbook' grammar, implemented by Christopher Callison-Burch, Scott Guffey and Dan Flickinger, based on Sag and Wasow (1999). Aline Villavicencio provided the sample categorial grammar and the answer grammars distributed in conjunction with this book.

Many thanks to Aline Villavicencio for book debugging and assistance with production of the final version and also to Tony Gee at CSLI Publications. I am very grateful to Dikran Karagueuzian for his support and advice in multiple contexts, including those totally unrelated to book production.

The LKB system development is very much a collaborative effort, and I am truly fortunate to have such gifted and friendly collaborators — working together has been a remarkably enjoyable experience (at least from my perspective). I am deeply grateful to them and to my family for their support and tolerance.

Part I

Typed Feature Structure Grammars

1

Introduction

This book is an introduction to practical grammar implementation for people who are interested in linguistics and computational linguistics. Hand-built grammars are used in many natural language processing applications, either to analyse spoken or written input or to generate text or speech. Writing such grammars is effectively a complex programming task, one that requires some linguistic knowledge, in order to understand what the phenomena are, and some skill in developing an implementable account of the phenomena. Grammar writers also need to know the details of the formalism in which they will implement the grammar. A formalism is like a very high-level programming language, one specifically designed for grammar development. Similarly, a grammar development environment is like a programming language environment: it contains various tools to help the developer construct the grammar besides allowing the grammar to be 'run' as a parser or generator. So developing a computational grammar is often called *grammar engineering*.

There are three interrelated aims of this book:

1. To give a gentle but precise account of one particular grammar formalism: a version of typed feature structures.[3]

2. To provide a hands-on introduction to the use of that formalism within an open source grammar development environment (the *Linguistic Knowledge Building* system, generally known as the *LKB system* or simply the *LKB*).

3. To illustrate some of the techniques of grammar engineering by providing a detailed account of some sample grammars.

These aspects are interwoven in the first half of the book: the formalism is motivated by the grammar encoding and the user is encouraged to try

---

[3]Typed feature structure formalisms are a variety of *unification-based* formalism. Such languages are also often described as *constraint-based*.

things out in the LKB system.

The book is intended to be accessible to people who do not like equations, mathematical symbols, Greek letters and so on. Formal definitions are given, but only after a precise non-mathematical description of the concept, so it is never necessary to understand the formal definitions to follow what is going on. Many exercises are suggested: answers are provided to most of them.

This book is also a reference manual for the LKB system. The second half of the book describes the user interface, details the error messages and the system parameters. It also contains brief details of various aspects of the system which are relatively advanced.

I have tried hard not to 'dumb down' the non-formal descriptions of the LKB formalism: they are intended to be accurate and complete, not simplified for pedagogical purposes. Grammar engineers need to know precisely what the system is supposed to do! However, the introductory nature of this book has meant that I have mostly restricted myself to describing the formalism used in the LKB system and have not made comparisons with other approaches. I have also avoided any discussion of formal complexity results: the remarks about efficiency made in the book are based on experience with implemented systems. However some references are given to books and research papers which do discuss these aspects.

After working through this book you will be able to write your own grammars for parsing and generation. In conjunction with more linguistically oriented works, this book should provide enough background for readers to understand most of the details of the existing large grammars which can run in the LKB. Such grammars have been used in real-life commercial applications, so although this book mostly discusses toy grammars, grammars can be built with the system that are practically useful. Although I will only discuss English here, grammars for at least seven other languages have been developed using the LKB.

1.1   What you need to run the LKB

This book is designed to be used in conjunction with the LKB system software. The LKB is available free of charge as open source software from the LKB website:

http://cslipublications.stanford.edu/lkb.html

The licence imposes no conditions on use of the LKB: but for full details please see the website.

The LKB system runs on Microsoft Windows (95, 98, NT, 2000), Linux and Sun Solaris.[4] You will need to be able to download it from the website: the size of the downloadable (compressed) files is around 3.5 Megabytes.[5] Download instructions are given in §2.1 and on the website. The fully expanded files require about 25 Megabytes of disk space. The LKB system is rather memory hungry when running: for small grammars, such as those in this book, I recommend a minimum of 128 Megabytes of memory. A large grammar, such as the LinGO English Resource Grammar (see §1.3), requires considerably more memory: we have found that 384 Megabytes gives satisfactory performance.

1.2 What you need to know about to understand this book

I assume that the reader has some idea of what a natural language grammar is and why it is useful. One good textbook which covers the relevant concepts is Jurafsky and Martin (2000). You should know, for instance, what a context-free grammar is (at least approximately — it doesn't matter much if you don't exactly remember details like the formal distinction between context-free and context-sensitive). A little knowledge of natural language parsing techniques will also be helpful, although this shouldn't be essential. The reader will need a very basic knowledge of set theory (e.g., the definition of *intersection, union* etc) and of logic (e.g., being able to read expressions such as $\exists x[dog(x) \land sleep(x)]$) — Allwood et al (1977) is a good introduction. Some sections will require rather more background than this for detailed understanding, but these sections are marked as not essential.

The discussion also assumes some relatively basic linguistic concepts and terminology (e.g., *syntax, semantics, lexicon, tree, noun phrase, agreement*). In order to build grammars of your own, you will need to be able to apply concepts such as tests for constituency. For instance, you need to understand why the following sentences are generally assumed to have different syntactic structures:

(1.1)   Kim looked up the chimney.

(1.2)   Kim looked up the answer.

Tallerman (1998) is an introductory book which covers all the concepts

---

[4]Running the LKB on an Apple Macintosh is possible but requires that additional software be purchased — this is discussed on the LKB website.

[5]Download from this website is the preferred form of distribution, since the LKB is under continual development and by downloading you get the most recent stable version of the system. But if you are unable to download large files, you may request a CD from the publisher by sending off the form at the back of the book.

I will assume relatively informally.

Sag and Wasow (1999) is an introduction to syntax which uses a formalism quite similar to the one described here. Some of the grammars provided with the LKB system are implementations of the grammars described by Sag and Wasow. But it is not necessary to have worked through Sag and Wasow's book before reading this one, and in fact anyone who has done so needs to be aware that there are substantial formal and terminological differences, which will be discussed in some detail later in this book.

It is necessary to run the LKB software to get the most out of this book, and a certain degree of familiarity with computers is assumed. I have given quite detailed instructions for installing and running the LKB (see Chapter 2) but you will have to know how to do things like open and resize windows on your system, select items from a menu, open a file and so on. You also need to be able to use an editor suitable for writing code rather than text. For Linux, Solaris and Windows, I recommend *emacs* (either gnuemacs or Xemacs) — details of how to download emacs and use it with the LKB system are given on the LKB website.

1.3 A brief introduction to the LKB system

The LKB system is a grammar and lexicon development environment for typed feature structures grammars. It has been most extensively tested with grammars based on Head-Driven Phrase Structure Grammar (HPSG: Pollard and Sag, 1987, 1994), but it is framework independent. In this philosophy, and in much else, it follows PATR (Shieber, 1986). The LKB source is freely available: the system is implemented in Common Lisp and the basic code is intended to run on any ANSI-compliant Common Lisp implementation. The graphical environment is currently limited to Allegro Common Lisp with the Common Lisp Interface Manager (CLIM), although a subset of the commands are available for Macintosh Common Lisp.[6]

The best way to think about the LKB system, and other comparable systems (see §5.8.2) is as a development environment for a very high-level specialized programming language. Typed feature structure languages are essentially based on one data structure — the typed feature structure, and one operation — unification. The type system constrains the allowable structures and provides a way of capturing linguistic generalizations. This combination is powerful enough to allow the grammar

---

[6] **Allegro CL** is a trademark of Franz Inc. **MCL** and **Macintosh Common Lisp** are trademarks of Digitool, Inc. **Macintosh** is a trademark of Apple Computer. All other trademarks used in this book are the property of their owners.

developer to write grammars and lexicons that can be used to parse and generate natural languages. In effect, the grammar developer is a programmer, and the grammars and lexicons comprise code to be run by the system.[7] But because the typed feature structure language is so high level, working in it requires relatively little knowledge of computers. Typed feature structure languages were generally designed as a formal way of specifying linguistic behaviour rather than as programming languages, and they are used in this way by people who have no involvement in computational linguistics. So, at least potentially, these languages have several advantages: they make computational linguistics and natural language processing accessible to linguists with a very limited background in computer science, they enable computational linguists to adopt techniques from theoretical linguistics with minimal reinterpretation, and they allow formal theories to be tested on a range of data in a way that allows the interaction between phenomena to be checked.

The LKB system is a software package for writing linguistic programs, i.e., grammars, and it's important not to confuse the system and the grammars which run on it. By itself the LKB system is of no use in an application, since there has to be some associated grammar. Although some grammars are distributed with the LKB, most of them are just there to act as examples. The LKB system can be used to develop many different grammars. What's more, a grammar developed on the LKB system could be run on another platform that used the same typed feature structure language, just as a C++ program can be run by a variety of compilers.[8] The performance of the LKB system in tasks such as parsing depends critically on the grammar: it makes little sense to talk about performance except with respect to a particular grammar. We have, however, tried to make it reasonably efficient with fairly large grammars.

The largest grammar that has so far been developed on the LKB system is the LinGO project's English Resource Grammar (ERG: see Copestake and Flickinger, 2000 and `http://lingo.stanford.edu`). The LinGO ERG runs on several systems beside the LKB, including

---

[7]Throughout this book, I'll use grammar to mean all the linguistic code: that is grammar rules, lexicon, lexical rules and so on.

[8]In practise, however, there is no formally agreed standard for typed feature structure systems. The current situation is that some grammars can be converted relatively easily between platforms, while others are extremely difficult to convert because they use facilities which are not shared by all typed feature structure based systems. Developers of the different systems are trying to make portability easier, so the situation should continue to improve. Several systems can now run the LinGO English Resource Grammar (see below), with varying degrees of preprocessing.

PAGE (previously known as DISCO, Uszkoreit et al, 1994; Oepen et al, 1997), on which the initial ERG development was done. The ERG is being distributed in parallel with the LKB system, and download instructions for it are on the LKB website, but it is not discussed in detail in this book.

The LKB system and the ERG are both under active development, as a collaborative effort between researchers in several different countries. Some of the papers published about this research are cited in this book, but the LinGO and LKB websites should be consulted for further references and for up-to-date details.

1.4   Using this book

Although I hope this book will be useful on formally taught courses, I have tried to make it accessible to people who are working on their own and have some knowledge of linguistics and computational linguistics, as discussed above.

The book is divided into two parts. The first part is the introduction to grammar engineering and the typed feature structure formalism using the LKB, while the second is purely a reference for the LKB system. Chapter 2 is designed as a tour of the LKB for first time users: it illustrates many of the functions that will be seen in much more detail in the rest of the book. It uses a small but fairly complex grammar: the subsequent chapters explain everything illustrated in this chapter. Chapter 3 describes the typed feature structure formalism in precise detail, illustrating the concepts with very tiny grammars. Chapter 4 moves on to talking about parsing and generation with typed feature structures, both in theoretical terms and as implemented in the LKB system. It also specifies the way that grammars are written using the LKB's description language. At this point, the reader should understand all the formal issues that underly the LKB and have a good idea of how simple grammars are put together. Chapter 5 goes on to talk about more linguistically interesting grammars, describes lexical and morphological rules and contains an illustration of how simple semantics can be encoded. Suggestions for further reading are at the end of that chapter. Chapters 3, 4 and 5 contain a range of exercises: answers to nearly all of these are given in the text or as grammars that can be downloaded from the LKB website.

The chapters in the second half of the book are designed to act as a reference manual for the LKB system — they are not intended to be read straight through, but to act as a source of information about error messages, user interface commands which have less than obvious behaviour and so on. Readers who work through the exercises in the first half of the book will almost certainly need to refer to some of these chapters when stuck. Chapter 6 covers the user interface, Chapter 7 describes error messages and discusses some tools which are useful for debugging. Chapter 8 covers 'advanced' features, while Chapter 9 gives details of the user-settable system parameters: chapters 8 and 9 are likely only to be relevant to users who go on to develop their own grammars.

2

A first session with the LKB system

The following chapter takes the new user through an initial session with the LKB system. It covers the basics of:

1. Obtaining and starting the LKB
2. Using the LKB top menu
3. Loading an existing grammar
4. Examining typed feature structures and type constraints
5. Parsing sentences
6. Viewing a semantic representation
7. Generating from parse results
8. Adding a lexical entry
9. Adding a type with a constraint description

If you have no previous exposure to typed feature structure formalisms you will find that you don't fully understand all the terminology and notation used here. In the subsequent chapters, I will go through a sequence of grammars, starting with a very simple one and finishing off with the one that is illustrated in this chapter, explaining all the details of the formalism, so that you end up with a full understanding of how everything works. The idea is that the very detailed information will be easier to digest after this quick guided tour has given you an intuitive idea of where we are going.

2.1 Obtaining and starting the LKB

The instructions in this section outline how to get an executable version of the LKB for Windows, Linux or Solaris. Because the details may change slightly, you should also refer to the instructions for downloading on the LKB website:

http://cslipublications.stanford.edu/lkb.html

In case of any conflicts, follow the instructions given there rather than the ones in this book.

2.1.1 System requirements

Windows systems

1. Windows 95, 98, NT or 2000. The system may run on other versions of Windows but this has not been tested at the time of writing (check the website for updates).
2. At least 128 Megabytes of memory.
3. 30 Megabytes of free disk space.
4. WinZip or PowerArchiver for extraction of the downloaded files.

Linux

1. The system has been extensively used with Red Hat Linux (6.0 and later). Other versions should also work but have not been tested (check the website for updates).
2. At least 128 Megabytes of memory.
3. 30 Megabytes of disk space.
4. A suitable version of Motif: currently the LKB works with Metro Link Incorporated's Metro Motif 1.2.4 and OpenMotif. The LKB does NOT work with lesstif at the time of writing.
5. gzip and tar are needed to extract the files.

Solaris Solaris requirements are similar to Linux, but Motif is generally already installed on Solaris systems. OpenMotif is not available for Solaris.

emacs We recommend the use of emacs (either gnuemacs or XEmacs) with the LKB for editing the grammar files but this is not essential, and it is not needed for this chapter. Instructions for obtaining emacs and setting it up with the LKB are given on the website.

2.1.2 Downloads

You will need to download two archives of files from the LKB website:

http://cslipublications.stanford.edu/lkb.html

One is the executable version of the LKB for whichever platform you intend to use, the other is a collection of example grammars.

Before downloading anything, I suggest that you make a new directory/folder[9] for all the LKB files. I will refer to this as your LKB directory.

---

[9]A *directory* is the equivalent in the Linux/Unix world to a *folder* in Windows/Macintosh terminology. I will use the term directory rather than folder throughout the rest of this book. I will also generally use the Linux/Unix notation,

Use your browser to locate the relevant version of the LKB from the LKB website and save the file to your LKB directory. You will then have to extract the compressed files: on Linux or Solaris you can use `gzip` followed by `tar xf` while on Windows, WinZip or PowerArchiver should uncompress and extract as one operation.

You can then download and extract the data files into the same directory. This should result in a directory `data`, with a number of sub-directories, including `itfs`, which is the directory for all the grammars we will use in this book.

Instructions for building the LKB from source files are on the website.

2.1.3 Starting the LKB

Before you start the LKB for the first time, you must create a temporary directory/folder. On Linux or Solaris, this should be a directory called `tmp` in your home directory. On Windows, create the empty folder `C:\tmp`. If necessary, the location of this directory can be varied, details of how to do this are given on the LKB website.

To start the LKB on Windows, simply double click on `lkb.exe`. On Linux or Solaris, cd to your lkb directory and type `lkb` at the command line. If you have successfully started the LKB, you should see the LKB top menu window, as described in the next section. You will also see a command line, with prompts such as `LKB(1):` — this allows the user to type in commands but can be ignored for the purposes of this chapter.

Warning note: do not use the NumLock key when using the LKB system with Linux. There is a bug in the software on which the LKB is built which causes menus to stop working intermittently when the NumLock key is on. Once this has happened, restarting the LKB itself will not help, you have to log out and restart your X session.

2.1.4 Installation problems

In case of installation or other problems which don't seem to be covered in the documentation, please look at the instructions on the LKB website. We will make fixes for known problems available there. The website also contains details of how to report any bugs that you find.

2.2 Using the LKB top menu

.

The main way of interacting with the LKB is through the LKB top menu window which is displayed once you have started the LKB, as shown below.

---

with forward slash (/) separating directories, rather than the Windows notation, but the equivalence should be obvious.

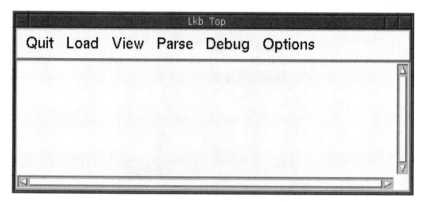

This is a general purpose top level interaction window, with the menu displayed across the top — most LKB system messages appear in the pane below the menu buttons.[10] I will use the term *LKB interaction window* for the window in which messages appear. Most of the menu commands are not available when the LKB is first started up, because no grammar has been loaded.

2.3 Loading an existing grammar

The first step in this guided tour is to load an existing grammar: i.e., a set of files containing types and constraints, lexical entries, grammar rules, lexical rules, morphological rules and ancillary information. The LKB comes supplied with a series of grammars: the ones we will use in this book are all in the directory `data/itfs/grammars`. In this section, we will assume that you are working with the grammar in `g8gap`.

To load a grammar it is necessary to select a *script* file which controls how the grammar files are loaded into the system. Select **Complete grammar** from the LKB **Load** menu, and choose the file `script` from the `g8gap` directory as shown below.

---

[10]The Macintosh version of the LKB has a slightly different user interface which I will not describe in this book: for a brief discussion of the main differences, see the LKB web page.

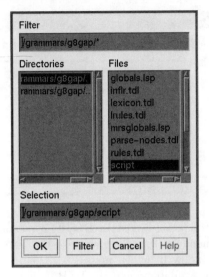

You should see various messages appearing in the interaction window, as shown in Figure 1 (the interaction window has been enlarged). If there are any errors in the grammar which the system can detect at this point, error messages will be displayed in this window. With this grammar, there should be no errors, unless there is a problem associated with the temporary directory (see §2.1). If you get an error message when trying to load **g8gap/script**, it is possible you have selected another file instead of **script** — try again.[11]

Once a file is successfully loaded, the menu commands are all available and a *type hierarchy* window is displayed (as shown below). You can enlarge this window to show the complete hierarchy or scroll it in the usual way.

---

[11]In case of genuine problems, please see the LKB webpage section on known bugs and bug reporting.

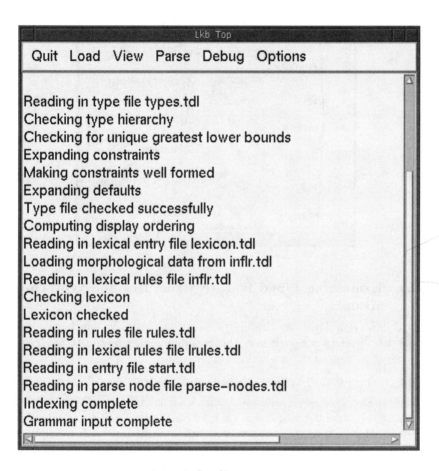

FIGURE 1 Loading a grammar

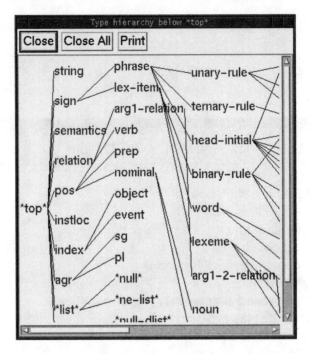

2.4 Examining typed feature structures and type constraints

In this section I will go through some of the ways in which you can look at the data structures in the grammar, such as the types, type constraints, lexical entries and grammar rules.

2.4.1 The type hierarchy window

The backbone of any grammar in the LKB is the type system, which consists of a hierarchy of types, each of which has a *constraint* which is expressed as a *typed feature structure*. Constraints are used to capture generalizations: the type hierarchy allows for inheritance of constraints. In the LKB system, the type hierarchy window is shown with the most general type displayed at the left of the window.[12] In this grammar, as in all the grammars I will discuss, the most general type is called ***top***. You will notice that there is some *multiple inheritance* in the hierarchy

---

[12]In the book, except where showing a screen dump, I will show type hierarchies with most general type towards the top of the page: the reason for the alternative orientation in the LKB itself is that this is a more efficient layout for typical hierarchies, which tend to be broad rather than deep.

(i.e., some types have more than one parent). You will see a few types with names such as **glbtype1**: these are types which are automatically created by the system, for reasons which will be explained in the next chapter.

Click on the type ***ne-list*** which is a daughter of ***list***, which is a daughter of ***top***, and choose **Expanded Type** from the menu. A window will appear as shown below.

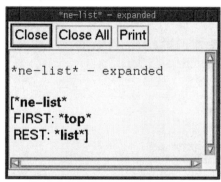

This window shows the constraint on type ***ne-list***: the constraint is expressed as a typed feature structure (TFS). It has two *features*, FIRST and REST. The value of FIRST is ***top***, which indicates it can unify with any TFS since ***top*** is the most general type. The value of REST is ***list*** which indicates it can only unify with something which is of type ***list*** or one of its subtypes. ***list*** and its daughters are important because they are used to implement list structures which are found in several places in the grammar. In this book, and in the LKB system windows, types are shown in lowercase, bold font, while features are shown in uppercase (small capitals in the book).

Look at the entry for the type ***ne-list*** in the actual source file g8gap/types.tdl, by opening that file in your editor (if you are using Windows, and have not installed emacs, you should use Notepad). If you search for *ne-list*, you will see the following definition:

```
*ne-list* := *list* &
 [ FIRST *top*,
   REST *list* ].
```

The language in which the type and its constraint are defined in the files is referred to as a *description language*. The type definition must specify the parent or parents of a type (in this case, ***list***) and optionally gives a constraint definition. In this particular case, the constraint described in the file corresponds very closely to the expanded constraint shown in the typed feature structure window, because the only parent of ***ne-list*** is ***list*** and this does not have any features in its constraint. However, in general, type constraints inherit a lot of information from the ancestors of the type, so the description of a constraint is usually very compact compared to the expanded constraint.

To see a more complicated type constraint, click on **phrase** in the type hierarchy window (found via **sign** from ***top***) and again choose **Expanded Type**. The TFS window is shown below:

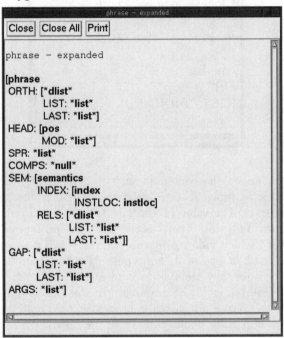

This illustrates that types can have complex constraints — the value of a feature in a constraint can be a TFS.

Look at the definition of **phrase** in the source file g8gap/types.tdl:

```
phrase := sign &
  [ COMPS < > ].
```

The notation < > is an abbreviation for ***null***, which represents a list with no elements. In contrast to ***ne-list***, the expanded constraint on **phrase** has inherited a lot of information from other types in the hierarchy: you can get an idea of how this inheritance operates by looking at the constraint of **phrase**'s parent (**sign**) in the type hierarchy window. Note that **sign** has a feature SEM which has the value **semantics**: some of the information in the expanded constraint on **phrase** comes from the constraint on **semantics**.

You will find that you can click on the types within the TFSs to get menus and also on the description at the top of the window (e.g., `phrase - expanded`). I won't go through all these menu items here, however, but they are discussed in Chapter 6.

2.4.2 The View commands

The view commands let you see objects such as *lexical entries* which are not types and therefore cannot be accessed from the type hierarchy window.

Select **View** from the LKB top menu and then select **Word entries**. You will be prompted for a word which corresponds to a lexical entry. Enter `dog` (case doesn't matter), deleting the default that is specified (unless of course the default is `dog`, in which case just select OK). You should get a TFS window corresponding to the entry which has orthography `"dog"` in the `g8gap` grammar as shown in Figure 2. If there were multiple entries with the spelling `"dog"` they would all be displayed.

You should compare the TFS shown in the window with the lexical description in the file `g8gap/lexicon.tdl`, to see how the inheritance from type constraints operates. The lexical description for *dog* is simply:

```
dog := noun-lxm &
[ ORTH.LIST.FIRST "dog",
  SEM.RELS.LIST.FIRST.PRED "dog_rel" ].
```

Nearly all the detail in the full TFS comes from the type **noun-lxm**.

Now try **View Grammar rule** and enter `head-specifier-rule` (or choose it from the selections if a menu is displayed). You will see that the grammar rule is also a TFS, which is shown in Figure 3. I do not reproduce it in full here, because it will not fit on one page, but the boxes indicate which parts of the structure have been 'shrunk' (this can be done by clicking on a node in a TFS window, and choosing the menu option **Shrink/expand**). A TFS that encodes a rule can be thought of as consisting of a number of 'slots', into which the phrases for the daughters and the mother fit. In this grammar, as in all the others we will look at in this book, the mother is the TFS as a whole, while the daughters are the elements in the list which is the value of the ARGS feature.

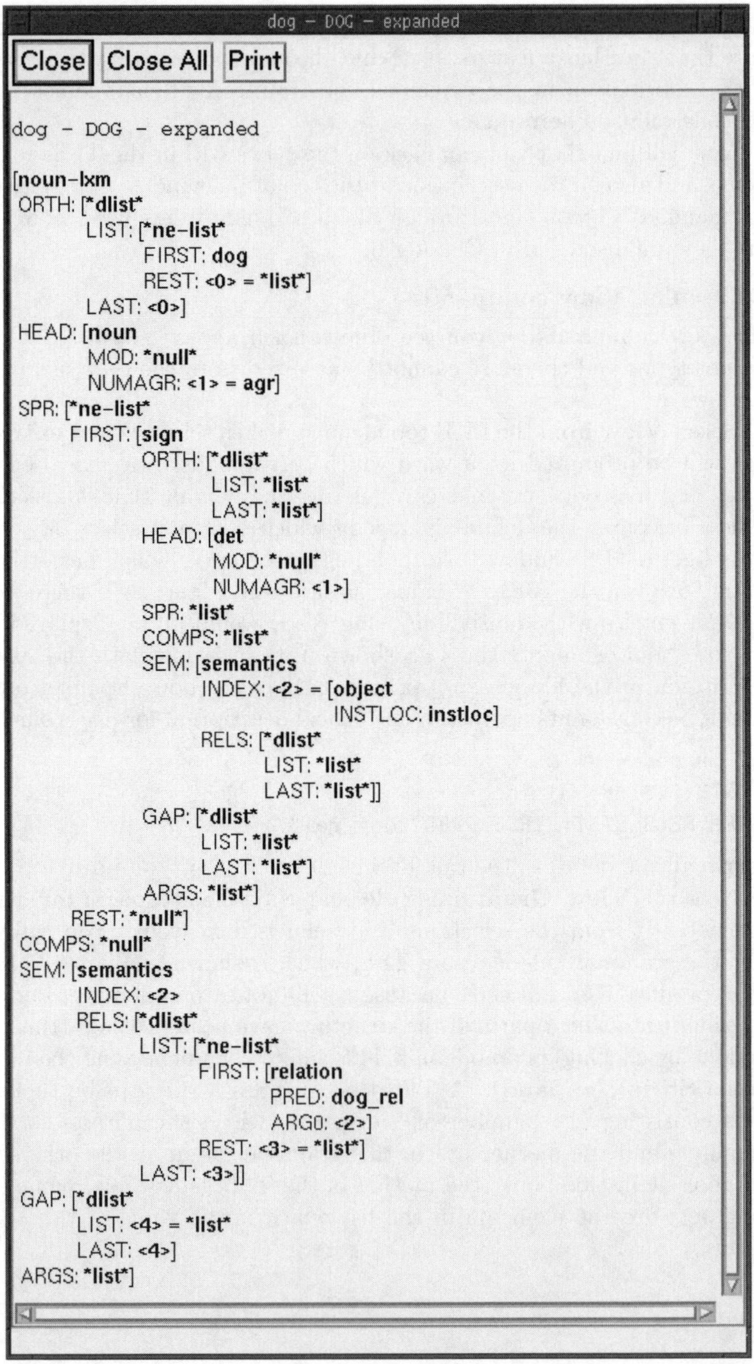
dog - DOG - expanded
Close
Close All
Print
dog - DOG - expanded
[noun-lxm
ORTH: [*dlist*
LIST: [*ne-list*
FIRST: dog
REST: <0> = *list*]
LAST: <0>]
HEAD: [noun
MOD: *null*
NUMAGR: <1> = agr]
SPR: [*ne-list*
FIRST: [sign
ORTH: [*dlist*
LIST: *list*
LAST: *list*]
HEAD: [det
MOD: *null*
NUMAGR: <1>]
SPR: *list*
COMPS: *list*
SEM: [semantics
INDEX: <2> = [object
INSTLOC: instloc]
RELS: [*dlist*
LIST: *list*
LAST: *list*]]
GAP: [*dlist*
LIST: *list*
LAST: *list*]
ARGS: *list*]
REST: *null*]
COMPS: *null*
SEM: [semantics
INDEX: <2>
RELS: [*dlist*
LIST: [*ne-list*
FIRST: [relation
PRED: dog_rel
ARG0: <2>]
REST: <3> = *list*]
LAST: <3>]]
GAP: [*dlist*
LIST: <4> = *list*
LAST: <4>]
ARGS: *list*]

FIGURE 2 Expanded lexical entry for *dog*

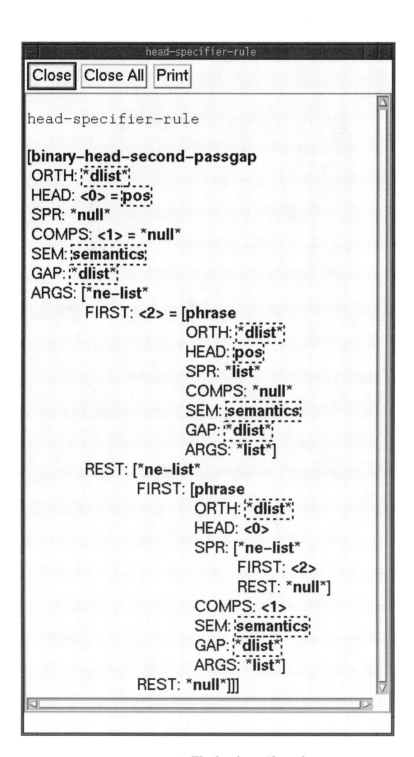

FIGURE 3 The head specifier rule

2.5 Parsing sentences

To parse a sentence, click on **Parse / Parse input**. A suitable sentence to enter is the dog barks. Click OK to start parsing. You will get a window with one tiny parse tree, as shown below.[13]

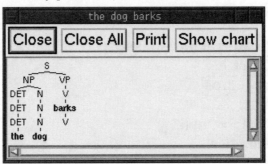

Although this grammar illustrates some quite complicated linguistic phenomena, it has a really tiny lexicon, so you can't type in arbitrary sentences and expect them to parse. To get an idea of what the grammar will parse, look at the test.items file.

2.5.1 Parse trees

If you click on the tiny parse tree in the window that shows the parse results, you will get a menu with an option **Show enlarged tree**. If you choose this, you will see a window with a more readable version of the tree, as shown below.

---

[13]The reason that the interface uses such a small size is that with non-trivial grammars and sentences, the parse trees can be very large and numerous, so this display is designed to allow a succinct overview.

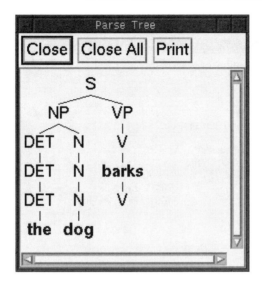

In the LKB system, a parse tree is just a convenient user interface device, which is shorthand for a much larger TFS. Click on the uppermost (S) node of the enlarged parse tree and choose the option **Feature structure — Edge 11**. You will see a large TFS, which is shown in Figure 4. As before, I have 'shrunk' some parts of the structure so that it can be displayed on the page. This structure represents the entire sentence. It is actually an instantiation of the `head-specifier-rule` shown in Figure 3.

The top node in the parse tree corresponds to the root node of the TFS shown in Figure 4. The structure for the phrase *the dog* is the node which is the value of the path ARGS.FIRST (a *path* is a sequence of features). The structure for the verb *barks* is the value of the path ARGS.FIRST.REST. The parse trees are created from the TFSs by matching these substructures against a set of node specifiers defined in the file `parse-nodes.tdl`. I will go into a lot more detail about how grammar rules work in the next chapters.

2.5.2 Morphological and lexical rules

You will notice that the parse tree has two nodes labelled V, one above and one below **barks**. They represent the application of a morphological rule: the lexicon contains an entry with the spelling `"bark"`, and the rule for third person singular verbs generates the inflected form `"barks"` from the lexical form. Morphological rules are used for inflectional and derivational processes which are associated with affixation: lexical rules

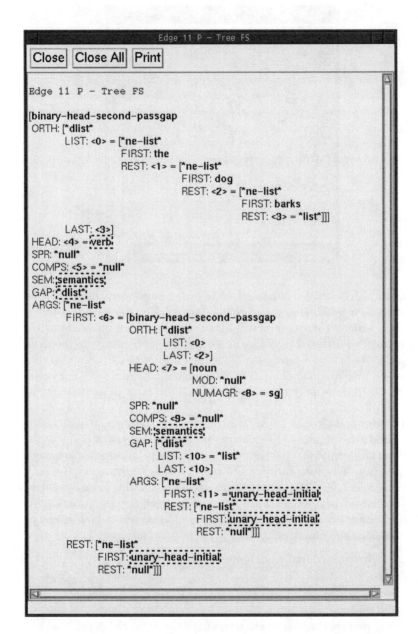

FIGURE 4 A TFS representing the sentence *the dog barks*

are used where there is no affixation. Morphological and lexical rules are very similar to ordinary grammar rules in the LKB system.

Try viewing the lexical entry for *bark* via **View / Lex entry**. This behaves much like **View / Word entries** that you used before, because in this small grammar, the identifiers for the lexical entries are the same as the orthography. Hence when prompted for a Lex-id, you can just enter bark. At the top left of the window, it will say bark - expanded. If you click on this, you will get a menu, which among other things has the option **Apply all lex rules**. If you select this, you will get a window which shows which lexical rules apply to *bark*: you can click on the nodes in the result window to display the feature structures corresponding to the inflected forms.

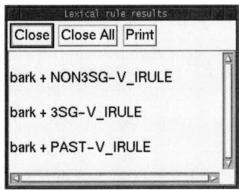

2.5.3 Batch parsing

Now try **Parse / Batch Parse**. You will be prompted for the name of a file which contains a test suite: i.e., a list of sentences which either should or should not parse in a grammar. A suitable file, test.items, already exists in the **g8gap** directory. Select test.items and enter the name of a new file for the output, e.g., test.items.out. The system will now parse all the sentences in test.items (this should only take a few seconds, unless you are using a very slow machine or one with very little memory). When you open the output file in your editor, it will show the following for each sentence:

1. the number of the sentence
2. the sentence itself
3. the number of parses (0 if there were no parses)
4. the number of passive edges (roughly speaking a *passive edge* represents a phrase that the system constructed while attempting to

parse a sentence)

For instance:

```
1 The dog barks. 1 11
2 *The dog bark. 0 10
```

Note that ungrammatical sentences are marked with an asterisk in the test suite file, and though a preprocessor strips off the asterisk and any punctutation symbols before attempting to parse, the results file shows the sentence in the form in which it was input. At the end of the results file, the total parsing time for all the sentences in the file is reported: of course, this time will depend on what sort of machine you have.

Have a look at the sentences in `test.items` to get some idea of the coverage of the grammar. You should try parsing some of these sentences individually and looking at the trees that result.

2.6 Viewing a semantic representation

Try parsing *this dog chased that cat*, and choosing **Indexed MRS** from the menu that you get by clicking on the small parse tree. This will give you a representation for the semantics of the sentence. (MRS is a semantic representation language that can be converted into more familiar languages such as predicate calculus, as is explained in more detail in later chapters.)

The actual semantics is constructed as a TFS, and can be seen as the value of the feature SEMANTICS in the TFS representing the parse for the sentence. The representation that is displayed when you select **Indexed MRS** is much easier to read, although some information has been omitted. A representation which is closer to the TFS structures can be obtained by choosing **MRS** instead of Indexed MRS from the menu associated with the small parse tree.

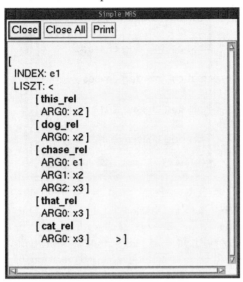

Semantic structures like this can be conveniently passed to other programs which use the results of the parse. One of the other options on the menu is **Prolog MRS**, which illustrates a format suitable for input to a Prolog system. (The **Scoped MRS** option which is also on the menu is not useful with this particular grammar, because it requires a fuller representation of quantifiers.)

2.7   Generating from parse results

The availability of semantics with this grammar allows us to try generating sentences. For convenience, this can be done simply by clicking on one of the small parse trees that results from parsing a sentence and choosing **Generate**. However, the actual input to the generator is the MRS representation, as shown in the windows you have just seen. Try generating from the result of parsing *this dog chases that cat*. The sentences generated are displayed in a window, as shown below.

You will notice that you get back four sentences. One is the input sentence and another is the present tense version of that sentence. The reason for getting both is that the semantic representation in this particular grammar does not include a representation of tense. You will also see two *topicalized* sentences: *that cat this dog chased* and *that cat this dog chases*. Topicalization may seem weird, but linguists generally assume that grammars should produce such sentences and in fact they are perfectly acceptable in some contexts.

If you click on one of the generated sentences, you will see a menu which has the options **Edge** and **Feature structure**. If you click **Edge**, you will see a tree for the sentence corresponding to the parse tree which would be produced if the sentence were parsed, although without any nodes corresponding to the inflectional rules.

Try parsing an ambiguous sentence, such as *this dog chased that cat near the aardvark* and compare the semantics for the different trees. Then generate from each tree. You should observe that you obtain a slightly different set of sentences. Some of these sentences will seem very weird, even if you have become used to topicalization! This grammar actually accepts/generates some ungrammatical sentences: this is what is formally referred to as *overgeneration*.

2.8   Adding a lexical entry

In this section, I will describe how to add a new lexical entry. You may want to make a backup copy of the g8gap directory before you start editing the files. First open the file **g8gap/lexicon.tdl** in your text editor. You will see that the grammar has three lexical entries for nouns — *cat*, *dog* and *aardvark*. Suppose you want to add another noun entry, perhaps for *rabbit*. This will look exactly like the entry for *dog*,

but with the value of the orthography path ORTH.LIST.FIRST replaced by
"rabbit" and the value of the semantic path SEM.RELS.LIST.FIRST.PRED
replaced with a suitable predicate, which I will call "rabbit_rel". The
"s round the orthography and semantic predicate values tell the system
that these are not 'proper' types: they are *string types* which do not need
to be declared. _rel, on the other hand, is just a naming convention:
the semantic predicate could equally well be "rabbit'" or "gavagai".
Add a new lexical entry to the lexicon file by copying and pasting the
entry for *dog* and changing dog to rabbit, as shown below:

```
rabbit := noun-lxm &
[ ORTH.LIST.FIRST "rabbit",
  SEM.RELS.LIST.FIRST.PRED "rabbit_rel" ].
```

Save the file, and then select **Load/ Reload grammar**. This will
reload the script file that you loaded before, this time loading the changed
version of the lexicon. You may get some error messages in the LKB interaction window at this point, if you have left out a character from the
new entry, for example. If you cannot see what is wrong, the description
of the error messages in §7.1.1 may help you track it down. When you
have successfully reloaded the system, you should be able to parse some
additional sentences, such as:

```
the rabbit chased the dog
```

Once you have successfully parsed some new sentences, you could add
them to the list of test sentences for batch parsing (i.e., test.items).

2.9   Adding a type with a constraint description

Substantial changes to the grammar always involve changing types. I will
illustate this with a very simple example. Consider nouns like *scissors*,
binoculars and *trousers*, which in most dialects of English always take
plural agreement. I'll refer to these as *pair nouns*. We don't want to give
such nouns a normal entry as a **noun-lxm**, because they would end up
behaving as standard nouns and having both singular and plural forms.
One very simple way to get approximately the correct behaviour with
this rather simple grammar is to make a new type which specifies that
the number agreement has to be plural. The instructions below walk
you through the process of creating the type: you are not expected to
understand exactly what is going on at this point, but just to get some
idea of how grammars may be modified.

Open the file **g8gap/types.tdl**. Search the file for the definition of
noun-lxm. Then add a new definition for **pair-noun-lxm** which should
inherit from **noun-lxm**, but specify that the value for HEAD.NUMAGR

is **pl**. The type definition you need to add is:

```
pair-noun-lxm := noun-lxm &
[ HEAD [ NUMAGR pl ]].
```

This specification can go anywhere in the file, though putting it under the definition of **noun-lxm** will improve readability,

You can then save the `types.tdl` file and check that you can load the revised grammar with **Reload grammar**. You should be able to see the new type in the type hierarchy and view its constraint. However, in order to demonstrate that the new type works as designed, we have to add a new entry to the lexicon file that uses it. As in the last section, you can do this by copying an existing entry, but this time you have to change the type to **pair-noun-lxm** as well as changing the orthography, the semantics and the identifier. For the sake of the example, we will enter the orthography as `"scissor"`, which will get the correct form when it has gone through the rule for plural noun inflection.[14]

```
scissor := pair-noun-lxm &
[ ORTH.LIST.FIRST "scissor",
  SEM.RELS.LIST.FIRST.PRED "scissor_rel" ].
```

Save the file and select **Reload grammar**. As before, check the LKB interaction window to make sure reloading was successful. Try parsing some new sentences. You should find that you cannot parse *the dog chased this scissor* but you can parse *the dog chased the scissors*.

Try generating from the result of parsing *the dog chased the scissors*. Note that the sentence with singular *scissor* is not generated. While accepting ungrammatical sentences is not necessarily too problematic for applications which are only intended to analyse sentences, we certainly don't want to generate them. One of the fundamental principles behind systems such as the LKB is that grammars can be *bidirectional*: that is, they can be used for both parsing and generation.

2.10 Summary

This tour has only touched on some of the features of the LKB system — there are several menu options which have not been described (these are all listed in Chapter 6). You should try playing around with the grammar, parsing and generating some more sentences, looking at how the TFSs are built up and adding a few more lexical entries that use the types already in the grammar.

The main aim of this chapter was to give you a rough idea of what

---

[14]This isn't unreasonable: a stem form *scissor* has to exist for some compound nouns, such as *scissor kick*, though we don't deal with compounds in this grammar.

can be done with a typed feature structure system and a simple grammar. Concepts such as types, typed feature structures, lexical entries, grammar rules, parsing, generation and semantics were introduced very briefly and informally. The next three chapters discuss all of this is full detail. Although the grammar in **g8gap** is very small, it does illustrate the main aspects of grammar engineering with typed feature structures and can be used as a basis for understanding or writing much larger grammars.

3

Typed feature structures made simple

If you have worked through the previous chapter, you should have gained an intuitive idea of the components of a typed feature structure grammar, how to write typed feature structure descriptions for lexical entries etc, and how grammar rules work. In this chapter I will give a full description of TFSs. The bulk of this chapter involves precise but informal accounts of the language used for writing grammars. Formal definitions are given as well, but it is not essential to understand these. This chapter could be read without access to a running LKB system, although having the system available will be necessary to work with the sample grammars and to do some of the exercises.[15]

Some sections in this chapter are followed by exercises, which it is important to attempt. They shouldn't take more than a few minutes and sometimes I will use them to introduce new material.

Before I start describing the details of the formalism, I am going to introduce a really tiny grammar, both in order to make the concept of encoding a grammar in TFSs clearer, and also to give concrete examples that can be used in the discussion of the TFSs themselves.

---

[15]Cautionary note: There are several slightly different variants of typed feature structure formalisms. Here I am only describing the version used in the LKB system, and since this is essentially an informal introduction, I will not discuss the differences with other approaches. The LKB TFS formalism is generally based on Carpenter (1992), but there are differences in the treatment of type constraints and well-formedness. The terminology in the literature is sometimes confusing and I have not tried to explain all the alternatives. However, I should note here that Sag and Wasow (1999) refer to typed feature structures as *typed feature structure descriptions*. They make a distinction between 'typed feature structure descriptions' and what they call 'typed feature structures' for which there is no simple counterpart in the formalism described here. However, the practical effect of these differences is somewhat limited, see §5.6.

3.1   A really really simple grammar

To start off with, I want to consider a very simple grammar, which can easily be defined in a conventional context free language. I expect that readers have some familiarity with grammars defined with atomic symbols (e.g., Backus-Naur notation). In 3.3 I show a very simple grammar and lexicon, with just two grammar rules and six lexical entries.

(3.3)   Start symbol: S   Lexicon:

        Rules: dog : N
                                     dogs: N
        S -> NP VP   this : Det
        NP -> Det N   these: Det
                                       sleeps : VP
                                       sleep: VP

The specification of the *start symbol* as S means that the grammar will accept as complete strings "these dogs sleep" and "this dog sleeps", but not "sleep" and "this dog", because these cannot be Ss by themselves according to this grammar. (In a larger grammar, the imperative "Sleep!" would be an S, but this is not allowed for here.)

Grammars in general can be viewed abstractly as mappings between a set of descriptive structures and a set of strings of characters. For a very simple context free grammar (CFG) like that in 3.3, we can use a labelled bracket notation as the description. For instance:

(3.4)   (S (NP (Det this) (N dog)) (VP sleeps))
         ↔ "this dog sleeps"

This labelled bracketing can equivalently be drawn as a tree, as shown in 3.5.

(3.5)

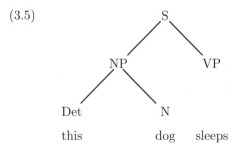

Grammars can also be viewed as specifications that allow parsing and generation of strings given partial information. For instance, given the input string:

```
these dogs sleep
```

we could parse it to derive the single labelled bracketing:

```
(S (NP (Det these) (N dogs)) (VP sleep))
```

Given the partial bracketed structure:

```
(S (NP (Det ?) (N ?)) (VP sleeps))
```

(where I am just using ? as a placeholder for unknown strings), the corresponding strings generated by the grammar are:

```
this dog sleeps
these dog sleeps
this dogs sleeps
these dogs sleeps
```

As this example shows, this grammar allows sentences which are not grammatical English, such as:

> these dog sleeps

The problem is that the grammar does not include any specification of agreement. It could be fixed by increasing the number of atomic symbols, for instance as shown in 3.6. But this approach would soon become very tedious as we expanded the grammar.

(3.6)   Start symbol: S   Lexicon:

```
         Rules:                           dog : N-sg
                                          dogs: N-pl
         S -> NP-sg VP-sg                 this : Det-sg
         S -> NP-pl VP-pl                 these: Det-pl
         NP-sg -> Det-sg N-sg             sleeps : VP-sg
         NP-pl -> Det-pl N-pl             sleep: VP-pl
```

Intuitively a generalization is being missed: ideally we want to say that the subject and verb phrase have consistent number values without actually saying what they are.

Feature structures are one way of allowing extra information to be encoded to deal with agreement and more complex phenomena. For instance, we could specify that for each phrase there is a structure representing it which contains the feature NUM which can have values **sg**, **pl** or **unspecified**. Some approaches to representation use such feature structures to augment a backbone of rules expressed using atomic symbols. However, in other approaches, including that implemented in the LKB, everything is a (typed) feature structure, including lexical entries and grammar rules.

Figure 5 shows how a grammar equivalent to the very simple grammar shown in 3.3 (i.e., the grammar without agreement) can be written using TFSs. This is the grammar that is in the directory `grammars/g1cfg` distributed with the LKB. You should load this grammar into the LKB system in order to experiment. You can load this grammar, and all the others I will discuss, in the same way as in §2.3: that is, select **Complete grammar** from the LKB **Load** menu, and choose the file `script` from the appropriate directory, which in this case is `grammars/g1cfg`. Figure 5 looks much more complex than the grammar given in 3.3, but the additional initial complexity of feature structure grammars pays off in wider-coverage grammars.[16]

I will not go through the operation of the grammar yet, but rely on the intuitive correspondence with the grammar in 3.3. For now, you should just note that the basic components of the grammar are:

The type system (most of left column of Figure 5)
> The type system acts as the defining framework for the rest of the grammar. For instance, it determines which structures are mutually compatible and which features can occur, and it sets up an inheritance system which allows generalizations to be expressed. For this particular grammar, for example, all the basic linguistic entities are of type **syn-struc** which the type system defines to have the feature CATEG. The information which was conveyed by the simple atomic categories in the grammar in 3.3 is associated with this feature. The categories themselves, **s**, **np** and so on, are also represented as types. All types must be explicitly defined, except for string types such as `"these"`.

Start structure The equivalent of the start symbol in the standard CFG in 3.3 is represented by the TFS at the bottom of the left column of Figure 5. Note that this is not a type, just a (named) TFS.

Lexical entries (most of right column of Figure 5)
> These are TFSs which the grammar writer constructs in order to encode the words of the language. Lexical entries define a relation-

---

[16]In fact this grammar is a bit more complex than it needed to be to simply simulate the grammar in 3.3 in the LKB system, but I've done this in order to be consistent with some of the other grammars distributed with the LKB, such as the grammar discussed in the previous chapter. As we'll see in a lot of detail later on, although the LKB system can process any grammar that can be expressed in this variant of the typed feature structure formalism, there are a few parameters which control the interface between the grammar and the system which have to be declared. For simplicity, I've kept most of those parameters the same in the sample grammars discussed in this book.

```
;;; Types                          ;;; Lexicon
syn-struc := *top* &               this := word &
[ CATEG cat ].                     [ ORTH "this",
                                     CATEG det ].
cat := *top*.
                                   these := word &
s := cat.                          [ ORTH "these",
                                     CATEG det ].
np := cat.
                                   sleep := word &
vp := cat.                         [ ORTH "sleep",
                                     CATEG vp ].
det := cat.
                                   sleeps := word &
n := cat.                          [ ORTH "sleeps",
                                     CATEG vp ].
phrase := syn-struc &
[ ARGS *list* ].                   dog := word &
                                   [ ORTH "dog",
word := syn-struc &                  CATEG n ].
[ ORTH string ].
                                   dogs := word &
string := *top*.                   [ ORTH "dogs",
                                     CATEG n ].
*list* := *top*.
                                   ;;; Rules
*ne-list* := *list* &              s_rule := phrase &
 [ FIRST *top*,                    [ CATEG s,
   REST *list* ].                    ARGS [ FIRST [ CATEG np ],
                                            REST [ FIRST [ CATEG vp ],
*null* := *list*.                                   REST *null* ]]] .

;;; Start structure               np_rule := phrase &
start := phrase &                  [ CATEG np,
[ CATEG s ].                         ARGS [ FIRST [ CATEG det ],
                                            REST [ FIRST [ CATEG n ],
                                                   REST *null* ]]] .
```

FIGURE 5 Tiny typed feature structure grammar: g1cfg

ship between a string representing the characters in a word and some linguistic description of the word (or, more precisely, of a particular sense of a word). In this grammar, the ORTH (orthography) feature is used to specify the string in lexical entries.

Grammar rules (bottom third of right column of Figure 5)

Grammar rules are TFSs that describe how to combine lexical entries and phrases to make further phrases. In the simple CFG, I expressed the idea of a rule in a conventional way by specifying a category for a phrase to the left of an arrow, and the daughter categories, in their expected linear order, to the right of the arrow. For this typed feature structure grammar the mother structure includes the daughters.[17] Thus, the rules are actually partial descriptions of phrases. The daughters of the phrase are contained in the feature ARGS. This takes as its value a list where the order of the list elements corresponds to the linear order of the daughters.

I will sometimes refer to start symbols, lexical entries and grammar rules collectively as *entries*.[18] Entries are ordinary TFSs (not types or type constraints) to which the grammar writer has given a name. Different classes of entries have different behaviour in the system.

In this book, types are always shown in lowercase bold font (e.g., **syn-struc**), features are always shown in small capitals (e.g., CATEG) and entries are always written with a 'typewriter' font (e.g., `s_rule`, `dogs`).

Intuitively, rule application in this grammar is much as in a CFG, but whereas in a CFG a rule matches a lexical entry or phrase if the category symbol is identical, in a typed feature structure grammar, the TFS in the relevant part of the rule needs to be compatible with the TFS for the lexical entry or phrase, where the two TFSs need not be identical in order to be compatible. What's more, information may be shared between the daughters and the mother of the rule. This is done via *unification* and much of this chapter involves describing precisely what the ideas of TFS compatability and unification mean.

For grammar `glcfg`, we could still derive the same labelled bracketing for *this dog sleeps* as we saw before, by simply taking the values of the category feature from the lexical entries and from each phrase. But with a feature structure grammar, a tree with atomic labels is just an

---

[17]This isn't a requirement of the LKB system but it's an approach I'll use in all the grammars considered in the book. The significance of this will become apparent later.

[18]By *entry* I mean roughly what was meant by *psort* in the old LKB documentation, and what is sometime referred to as an *instance*. I've changed the terminology because *psort* was just confusing and *instance* is somewhat inaccurate.

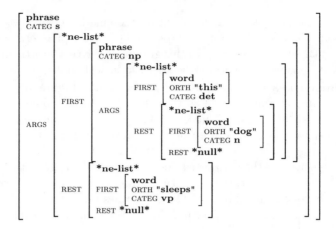

FIGURE 6 AVM for *this dog sleeps* from grammar **g1cfg** using TFSs

abbreviation for the complete representation. For this particular style of grammar, where the mother of a rule is represented as a TFS that contains all its daughters, the complete representation for the derivation is given by the TFS for the sentence. Such a structure is shown in Figure 6, represented as an *attribute-value matrix* (AVM).

In the next few sections, I will use examples from this grammar, and some slightly more complex variants of it, in order to explain TFSs in detail. I'll start off this detailed explanation of TFSs by discussing *type systems*, which form the backbone of grammars. A type system consists of a *type hierarchy*, which indicates specialisation and consistency of types, plus a set of *constraints* on types which determined which TFSs are *well-formed*. The next section describes the type hierarchy.

3.2   The type hierarchy

The diagram below (3.7) shows the type hierarchy corresponding to our simple grammar. I have just shown one string type (**"dog"**) explicitly, to illustrate how string types fit into the hierarchy. But all string types will be omitted in subsequent figures.

(3.7)

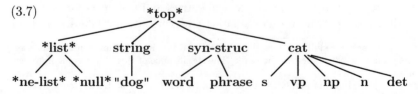

In all type hierarchies, there is a unique most general type, which I will refer to as the top type, and call ***top*** in all the grammars. The way that type files are written in the LKB means that the descriptions of types contain a specification of their parents together with the constraint (see Figure 5, where, for instance, ***ne-list*** is defined as having the parent ***list***). It is the specification of the parents that defines the type hierarchy. Note that ***top*** is not defined in the grammar specification. The hierarchy determines how constraints are inherited, although I won't get to the details of constraints until later in the chapter. The hierarchy is a tree, because no type has more than one parent, but it is possible for a type to have two or more parents: this is usually referred to as *multiple inheritance* and we'll see some examples of it shortly.

Below is a short summary of the properties of type hierarchies in general. (Later on, I'll give summaries of TFSs, constraints and so on.) All type hierarchies in the LKB system must obey these conditions.

Properties of type hierarchies

> **Unique top** There is a single hierarchy containing all the types with a unique top type.
>
> **No cycles** There are no cycles in the hierarchy.
>
> **Unique greatest lower bounds** Any two types in the hierarchy must either be incompatible, in which case they will not share any descendants, or they are compatible, in which case they must have a unique highest common descendant (referred to as the unique *greatest lower bound*).

In order to make the restrictions clearer, I will use artificial examples of hierarchies. Although I don't usually bother drawing arrows on hierarchies to show the direction, I will use vectors pointing from more general to more specific types in the next few figures to make things clearer.

The diagram in 3.8 shows a type hierarchy that violates the condition that there must be a unique top type (if you assume this is a single type system): here **animal** and **drink** are both top types.

(3.8)

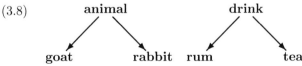

To make this valid, we would have to add a supertype as shown in 3.9.

(3.9)

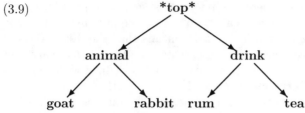

3.10 shows a hierarchy containing a cycle. This sort of situation only generally arises when a name is accidentally reused. I've used *primate* in this example, because the word is ambiguous between the class of animals that include apes and monkeys and a high-ranking bishop: therefore a primate (sense 2) is a primate (sense 1).

(3.10)

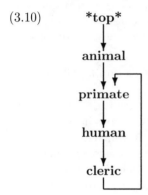

To fix this hierarchy, the concepts have to be renamed, e.g., as in 3.11.

(3.11)

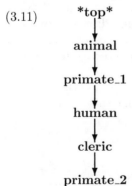

The most complicated property of the type hierarchy is the unique greatest lower bounds condition. To understand this, see 3.12.

(3.12)

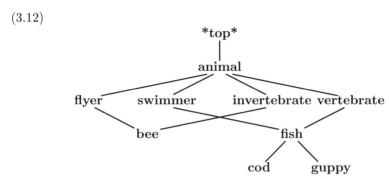

This is a formally valid hierarchy, since if we look at any pair of types, the relationship between them falls one of the following categories:

1. The types have no descendents in common (e.g., **vertebrate** and **invertebrate**)
2. The types have a hierarchical relationship (e.g., **animal** and **bee**), in which case the unique greatest common descendant is trivially the lower type.
3. There is a third type which is a unique greatest common descendant. For instance, **vertebrate** and **swimmer** have **fish** as a common descendant: **cod** and **guppy** are also common descendants, but **fish** is above both of them in the hierarchy.

Crucially, the assumption is made that all the types that exist have a specified position in the hierarchy (sometimes referred to as a *closed world* assumption) and that, if two types are compatible, there must be a single type which represents their combination. So if we know something is of type **vertebrate** and also of type **swimmer**, given this hierarchy, we can conclude it is of type **fish** (though we don't know whether it is a **cod** or a **guppy**).

An example of an invalid hierarchy is given in 3.13.

(3.13)

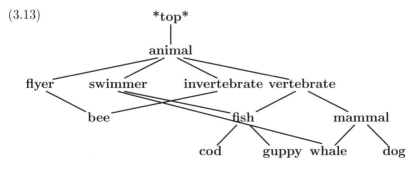

Here I have added **mammal** under **vertebrate** and **whale** as inheriting from **mammal** and **swimmer**. Now both **fish** and **whale** inherit from the types **vertebrate** and **swimmer**, but aren't related to each other through an inheritance relationship. This invalidates the hierarchy because it violates the unique greatest lower bounds condition. If **vertebrate** and **swimmer** are compatible there must be a single type for their combination, but both **whale** and **fish** independently inherit from them. Note that it is irrelevant that the node **mammal** intervenes between **vertebrate** and **whale** while **fish** is a direct daughter of **vertebrate**: **whale** is not a descendant of **fish**, so **fish** isn't the greatest lower bound of **vertebrate** and **swimmer**.

This hierarchy can be fixed straightforwardly, if somewhat uninterestingly, by adding a new node **vertebrate-swimmer**. This is shown in 3.14.

(3.14)

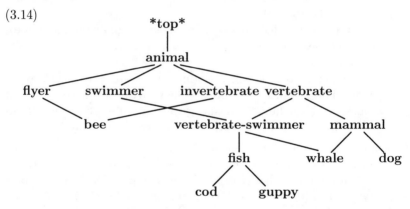

In the LKB, if a type hierarchy does not conform to the greatest lower bound (*glb*) condition, the system will automatically create types in order to satisfy the condition (these types are named **glbtype1**, **glbtype2** and so on and referred to as *glbtypes*). More details of this are given in §4.5.3. However, in this chapter, I will assume that all type hierarchies are specified by the grammar writer in such a way that they obey the glb condition.

Having described the restrictions on the type hierarchy, I should discuss the slight complication of strings. As mentioned above, strings such as `"these"` are an exception to the requirement that all types have to be explicitly defined by the grammar writer. Any arbitrary string (i.e., sequence of characters starting and ending with `"`) is conventionally considered to be a subtype of the type **string**. No strings have subtypes, which means that different strings are incompatible. So string types actually obey all the conditions on the type hierarchy, but they usually aren't shown in diagrams or in the type hierarchy display, since they have no interesting interrelationships.

You should note that, in the LKB system, despite the requirement that all types must be defined, it is valid to specify only one daughter for a type. For instance, suppose **count-n** were the unique daughter of **n**. The system would not infer that something of type **n** is necessarily of type **count-n**. This is one point of fundamental difference between typed feature structure formalisms: for instance, the approach assumed by Sag and Wasow would require that this inference be possible (as discussed in §5.6). The LKB system's form of the closed world assumption just requires the grammar writer to always specify, for any pair of types, whether or not they are compatible, and if they are compatible, what the result of combining them is. As we'll see below, this requirement makes it straightforward to define combination of TFSs, that is, unification.

3.2.1 *Formal definition of a type hierarchy

Here and below, the * in the subsection title indicates that this section can be skipped: the formal definitions only reiterate what has already been discussed.

A type hierarchy is a finite bounded complete partial order $\langle \mathsf{Type}, \sqsubseteq \rangle$.
The unique greatest lower bound of a set of types $S \subseteq \mathsf{TYPE}$ is written $\sqcap S$.
The maximal element $\top$ is defined so that $t \sqsubseteq \top$ for any t in TYPE.

3.2.2 Exercises

1. Draw the hierarchy which would correspond to the following descriptions (where x := a & b means that **x** has parents **a** and **b**):

   ```
   a := *top*.
   b := *top*.
   x := a & b.
   y := a & b & c.
   z := a & b & c.
   ```

   Why isn't this a valid type hierarchy? How could you fix it?

2. (Optional) Inheritance hierarchies are often introduced using examples from taxonomies. Biological taxonomies classify organisms according to categories such as phylum, class, order, family, genus, species and so on. In what respects are type hierarchies similar to and different from such biological taxonomies? Could a representation of a taxonomy be implemented using a type hierarchy?

3.2.3 Answers

1. The problems are:

   (a) Lack of connectivity: **c** is not defined.

   (b) Multiple greatest lower bounds

   In order to fix the lack of connectivity, we can define **c** to inherit from ***top***. In order to fix the multiple greatest lower bound problem, we have to introduce new types: a minimal solution is:

   ```
   a := *top*.
   b := *top*.
   c := *top*.
   ab := a & b.
   abc := ab & c.
   x := ab.
   y := abc.
   z := abc.
   ```

   If you came up with another solution, and want to check it in the LKB system, replace the type file in the directory grammars/an1 with your own type file and load the script in that directory as usual. If after you have loaded the grammar, the type hierarchy window does not show any types called **glbtype1** etc, your solution is acceptable. In general there is no unique way of adding types to turn a hierarchy into one that conforms to the greatest lower bound condition, although there is always a minimal solution.

2. Taxonomies and type hierarchies both encode a notion of specificity: for instance, the phylum *Chordata* contains the classes *Mammalia* and *Reptilia* (among others). They also share the property that the specificity relationship is transitive and cannot be overridden: there is no way to say that, for instance **bird** is a subtype of **vertebrate** and that **penguin** is a subtype of **bird** but to cancel the inference that **penguin** is a subtype of **vertebrate**. But there are differences — for one thing there is no cross-classification in conventional biological taxonomies, so they can be represented as trees. Furthermore, taxonomies use distinct levels of classification, although there are some categories which may not be instantiated for a particular organism, such as sub-species. So a taxonomy could be partially represented in a type hierarchy: the specialisation relationships could be captured, but the categories of classification, such as genus, cannot be represented, since there is nothing in the formalism which allows one to talk about the levels in the

hierarchy, or to restrict the hierarchy to a fixed number of levels. There's also something of a difference with respect to the notion of completeness: although organisms are discovered which can, for example, be classified with respect to genus but not to a particular species, a biologist is unlikely to be content to leave them underspecified for long. So it might be reasonable to assume that if there was only one species in a given genus, and that if an organism was classified as belonging to that genus, it automatically also belonged to that species. An LKB description of a taxonomy would also be deficient in this respect.

As with any knowledge representation problem, whether the deficiencies would actually matter depends on what the system is supposed to do!

3.3 Typed feature structures

I now move on to talking about typed feature structures. I'll make a distinction between TFSs in general, which I'll discuss in this section, and the subset of TFSs which are *well-formed* with respect to a set of type constraints, which are described in §3.5.

TFSs can be thought of as graphs, which have exactly one type on each node, and which have labelled arcs connecting nodes (except for the case of the simplest TFSs, which consist of a single node with a type but with no arcs). The labels on the arcs are referred to as *features*. Arcs are regarded as having a direction, conventionally regarded as pointing into the structure. For instance, in the structure in 3.15, there are two arcs, labelled with FIRST and REST, and three nodes, with types ***ne-list***, ***list*** and **word**.

(3.15)

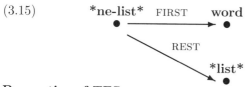

Properties of TFSs

> **Connectedness and unique root** A TFS must have a unique
> root node: apart from the root, all nodes have one or
> more parent nodes.
>
> **Unique features** Any node may have zero or more arcs
> leading out of it, but the label on each (that is, the
> feature) must be unique.
>
> **No cycles** No node may have an arc that points back to the
> root node or to a node that intervenes between it and

the root node.[19]

Types Each node must have a single type, which must be present in the type hierarchy.

Finiteness A TFS must have a finite number of nodes.

In 3.16, I show a more complex example which represents the structure for the rule np_rule from the grammar shown in Figure 5, where the first daughter position (ARGS FIRST) has been unified with the TFS corresponding to *these*. I will describe unification in detail in §3.4, but intuitively it just refers to combining two TFSs, retaining all the information in each of them.

(3.16)

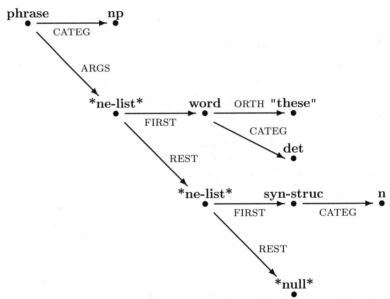

It is very important to realize that any non-root node in a TFS can be considered as the root of another TFS. For instance, consider the node reached by starting at the root of the structure in 3.16 then following the arc labelled ARGS and then the arc labelled FIRST. The TFS rooted at that node is shown in 3.17.

---

[19]Formally, in fact, both type hierarchies and TFSs are *directed acyclic graphs* (DAGs). However, directional arcs in TFSs do not encode specificity in any way. It would not make much sense for type hierarchies to contain cycles, since intuitively it cannot be the case that x is more specific than y and that y is also more specific than x. But since the arcs in TFSs don't have this sort of interpretation, cycles are not intuitively ruled out, and some variants of typed feature structure formalisms allow them.

(3.17)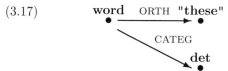

Because this graph notation is cumbersome (and difficult to draw in LaTeX ...), it is usual to illustrate TFSs with an alternative notation, known as an *attribute-value matrix* or AVM. The AVM corresponding to 3.16 is shown in 3.18.

(3.18)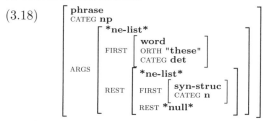

The *description language* used in the grammar files is similar to the AVM notation: a description which would correspond to 3.16 is shown in 3.19. The main differences are that the AVM notation we're using has the types inside the square brackets, while this description language puts them outside, and that the description language requires conjunction symbols &.[20] It is conventional to distinguish between features and types in descriptions by putting the former in uppercase, and the latter in lowercase, but case isn't formally significant. There is, however, nothing to prevent the same name being used for a type and for a feature, since the syntax of the descriptions always allows them to be distinguished.

(3.19)   example := phrase &
            [CATEG np,
              ARGS *ne-list* &
                    [FIRST word &
                            [ORTH "these",
                              CATEG det],
                      REST *ne-list* &
                            [FIRST syn-struc &
                                    [CATEG n],
                              REST *null*]]] .

Note that, although the description in 3.19 is perfectly valid syntacti-

---

[20]It would clearly be better if the description language mirrored the AVMs, but this is something we are stuck with for historical reasons, just like the QWERTY keyboard. The description language actually has a rather complex syntax that allows various abbreviatory notations. But discussion of this is left to §4.4.

cally, it's not usual to write something like this, both because there's no reason generally to write a description that corresponds to a partially instantiated rule, and because it makes explicit a lot of information that is inferred automatically via the process of making a structure well-formed, as we'll discuss in §3.5.

There is no significance to the order in which features are drawn. The following two AVMs represent exactly the same structure.

$$
\begin{bmatrix} \textbf{phrase} \\ \text{CATEG } \textbf{np} \\ \text{ARGS} \begin{bmatrix} \textbf{*ne-list*} \\ \text{FIRST } \textbf{word} \\ \text{REST } \textbf{*null*} \end{bmatrix} \end{bmatrix}
\qquad
\begin{bmatrix} \textbf{phrase} \\ \text{ARGS} \begin{bmatrix} \textbf{*ne-list*} \\ \text{REST } \textbf{*null*} \\ \text{FIRST } \textbf{word} \end{bmatrix} \\ \text{CATEG } \textbf{np} \end{bmatrix}
$$

However, it is obviously easier to read AVMs if the features appear in a consistent order (the way the LKB allows the user to define display order is described in §6.3 and §9.1.2).

It is often useful to talk about *paths* into feature structures: that is sequences of features that can be followed from the root node. In the example in 3.16, 3.18 and 3.19, for instance, the path ARGS.FIRST leads to a node with the type **word**. The *value* of a path is the TFS whose root is the node led to by the path (the TFS in 3.17 in this case). The root node in any TFS is the value of the *empty path*. I will sometimes use this terminology loosely however, and say that the value of a path is a type, when strictly speaking what I mean is that the value of the path is a TFS with a root node labelled with that type.

The graph in 3.16 is a tree: there are no cases where two arcs point to the same node. But *reentrancy*, as the non-tree situation is called, is required to build up more interesting typed feature structure grammars. 3.20 illustrates two TFSs: the first is not reentrant, while the second is. I will use structures with features like F and G here, to abstract away from the details of encoding in the grammar. (The types **t**, **a**, **b** and so on are assumed to all be defined in some hierarchy.)

If two paths point to the same node, we say the paths are *equivalent*. Reentrancy is also referred to as *coindexation* — it is conventionally indicated by boxed integers in the AVM diagrams and by tags beginning with **#** in the descriptions. The particular integer or tag used is of no significance: their only function is to indicate that the paths lead to the same node.

(3.20)

	Graph	AVM	description
Non-reentrant		$\begin{bmatrix} t \\ F\ a \\ G\ a \end{bmatrix}$	t & [F a, G a].
Reentrant		$\begin{bmatrix} t \\ F\ \boxed{0}\ a \\ G\ \boxed{0} \end{bmatrix}$	t & [F #1 & a, G #1].

To see reentrancy at work in a grammar, look at Figure 7, where I have augmented the grammar in Figure 5 with information about agreement (I have omitted some types from Figure 7 where they are unchanged from Figure 5). Try loading the grammar grammars/g2agr into the LKB system and parsing the sentences *this dog sleeps*, *these dog sleeps*. Look at the trees which result from the sentences which are admitted. The reentrancies which are specified on the rules ensure that:

1. the information about agreement in the lexical entries is passed up to the phrases
2. the determiner and noun have compatible agreement specifications in the rule np_rule
3. the noun phrase and the verb phrase have compatible agreement specifications in the rule s_rule

However, I can't fully describe this here, because I haven't discussed unification yet, so I will return to the details of how this works later.

```
;;; Types                              ;;; Lexicon continued

syn-struc := *top* &                   sleep := pl-word &
[ CATEG cat,                           [ ORTH "sleep",
  NUMAGR agr ].                          CATEG vp ].

agr := *top*.                          sleeps := sg-word &
                                       [ ORTH "sleeps",
sg := agr.                               CATEG vp ].

pl := agr.                             dog := sg-word &
                                       [ ORTH "dog",
phrase := syn-struc &                    CATEG n ].
[ ARGS *list* ].
                                       dogs := pl-word &
word := syn-struc &                    [ ORTH "dogs",
[ ORTH string ].                         CATEG n ].

sg-word := word &                      ;;; Rules
[ NUMAGR sg ].                         s_rule := phrase &
                                       [ CATEG s,
pl-word := word &                        NUMAGR #1,
[ NUMAGR pl ].                           ARGS [ FIRST [ CATEG np,
                                                       NUMAGR #1 ],
;;; Start structure                             REST [ FIRST [ CATEG vp,
start := phrase &                                            NUMAGR #1 ],
[ CATEG s ].                                          REST *null* ]]] .

;;; Lexicon                            np_rule := phrase &
this := sg-word &                      [ CATEG np,
[ ORTH "this",                           NUMAGR #1,
  CATEG det ].                           ARGS [ FIRST [ CATEG det,
                                                       NUMAGR #1 ],
these := pl-word &                              REST [ FIRST [ CATEG n,
[ ORTH "these",                                              NUMAGR #1 ],
  CATEG det ].                                         REST *null* ]]] .

the := word &
[ ORTH "the",
  CATEG det ].
```

FIGURE 7 Grammar with number agreement: g2agr

3.3.1 *Formal definition of a TFS

A TFS is defined on a finite set of features Feat and a type hierarchy $\langle \text{Type}, \sqsubseteq \rangle$. It is a tuple $\langle Q, r, \delta, \theta \rangle$, where:

1. Q is a finite set of nodes,
2. $r \in Q$ (r is the root node, see below)
3. $\theta : Q \longrightarrow \text{Type}$ is a total node typing function
4. $\delta : Q \times \text{Feat} \longrightarrow Q$ is a partial feature value function.

subject to the following conditions:

1. there is no node n or feature f such that $\delta(n, f) = r$ (root)
2. for every node n in Q there is a path π (i.e., a sequence of members of Feat) such that $\delta(r, \pi) = n$ (unique root, connectedness)
3. there is no node n or path π such that $\delta(n, \pi) = n$ (no cycles).

where I have extended δ to paths in the obvious way:

1. $\delta(q, \epsilon) = q$, where ϵ is the empty path
2. $\delta(q, f\pi) = \delta(\delta(f, q), \pi)$

This definition is taken fairly directly from Carpenter (1992).

3.3.2 Exercises

1. Which of the following are valid TFSs (assuming all the types used are defined)?

   (a) $\begin{bmatrix} \textbf{det} \end{bmatrix}$

   (b) $\begin{bmatrix} \textbf{t} \\ \text{F} \;\boxed{0} \\ \text{G} \;\boxed{0} \end{bmatrix}$

   (c) $\begin{bmatrix} \textbf{phrase} \\ \text{CATEG } \textbf{np} \\ \text{ORTH } \textbf{"these"} \\ \text{CATEG } \textbf{cat} \end{bmatrix}$

   (d) $\begin{bmatrix} \textbf{t} \\ \text{F} \;\boxed{0} \begin{bmatrix} \textbf{u} \\ \text{F} \;\boxed{1}\,\textbf{a} \\ \text{H} \;\boxed{1} \end{bmatrix} \\ \text{G} \;\boxed{1} \\ \text{J} \;\boxed{0} \end{bmatrix}$

   (e) $\begin{bmatrix} \textbf{t} \\ \text{F} \;\boxed{0} \begin{bmatrix} \textbf{u} \\ \text{F} \;\boxed{1}\,\textbf{a} \\ \text{H} \;\boxed{0} \end{bmatrix} \\ \text{G} \;\boxed{1} \\ \text{J} \;\boxed{0} \end{bmatrix}$

2. Draw the graph and the AVM for the structure for *this dog sleeps* in the version of the grammar with agreement (i.e., the grammar shown in Figure 7).

3. List all the valid paths in the following structure, and the types of the corresponding nodes. Also list the pairings of the path and the supertypes of the type on the node (assume **t**, **u** and **a** are immediate subtypes of ***top***). Also list all pairs of paths which lead to the same node.

$$
\begin{bmatrix} \mathbf{t} \\ \text{F } \boxed{0} \begin{bmatrix} \mathbf{u} \\ \text{H } \boxed{1}\,\mathbf{a} \\ \text{K } \boxed{1} \end{bmatrix} \\ \text{G } \boxed{1} \\ \text{J } \boxed{0} \end{bmatrix}
$$

3.3.3 Answers

1. (a) Valid. There are no arcs, but a TFS need only have a single node.

    (b) Valid, but only because of a notational convention which I haven't told you about! The structure looks invalid, since there is no explicit type associated with the inner node, and a type must be associated with every node in a valid structure. But there is a convention when writing AVMs that it is OK to omit the type on a node when showing a reentrant structure if the type is ***top***. So this structure is just shorthand for:

    $$
    \begin{bmatrix} \mathbf{t} \\ \text{F } \boxed{0}\,\text{*top*} \\ \text{G } \boxed{0} \end{bmatrix}
    $$

    (In fact, later on I will often omit the type on reentrant or non-terminal nodes when it can be inferred from the type constraints.)

    (c) Invalid. The structure has two features CATEG coming from the same node.

    (d) Valid. The reentrancy is fairly complex, but there are no cycles. Note that it is OK to have multiple cases of the feature F since they do not come from the same node.

    (e) Invalid. There is a cycle in the structure: the path F.H leads to the same node as the path F does.

2. Load the grammar in `grammars/g2agr` into the LKB, parse *this dog sleeps* and check the AVM against your answer. The graph you drew should have a single node in every place where the AVM shows a boxed integer. If you created the AVM using the LKB in the first place, award yourself one bonus point ...

3.

path	type
empty path	t
empty path	*top*
F	u
F	*top*
F.H	a
F.H	*top*
F.K	a
F.K	*top*
G	a
G	*top*
J	u
J	*top*
J.H	a
J.H	*top*
J.K	a
J.K	*top*

path	path
F	J
F.H	G
F.K	G
J.H	G
J.K	G
F.H	F.K
J.H	F.K
J.K	F.K
J.H	F.H
J.K	F.H
J.K	J.H

One can regard these pairings as comprising the individual pieces of information that are encapsulated in the single structure, though obviously there is some redundancy. Each of these pairs can be represented as a single TFS, as shown below (I have omitted the type ***top*** from the non-terminal nodes):

$$\begin{bmatrix} t \end{bmatrix} \quad \begin{bmatrix} G\ a \end{bmatrix} \quad \begin{bmatrix} F\ \boxed{1} \\ J\ \boxed{1} \end{bmatrix} \quad \begin{bmatrix} F \begin{bmatrix} K\ \boxed{1} \\ H\ \boxed{1} \end{bmatrix} \end{bmatrix}$$

$$\begin{bmatrix} *top* \end{bmatrix} \quad \begin{bmatrix} G\ *top* \end{bmatrix}$$

$$\begin{bmatrix} F\ u \end{bmatrix} \quad \begin{bmatrix} J\ u \end{bmatrix} \quad \begin{bmatrix} F \begin{bmatrix} H\ \boxed{1} \end{bmatrix} \\ G\ \boxed{1} \end{bmatrix} \quad \begin{bmatrix} F \begin{bmatrix} K\ \boxed{1} \end{bmatrix} \\ J \begin{bmatrix} K\ \boxed{1} \end{bmatrix} \end{bmatrix}$$

$$\begin{bmatrix} F\ *top* \end{bmatrix} \quad \begin{bmatrix} J\ *top* \end{bmatrix} \quad \begin{bmatrix} F \begin{bmatrix} K\ \boxed{1} \end{bmatrix} \\ G\ \boxed{1} \end{bmatrix} \quad \begin{bmatrix} F \begin{bmatrix} H\ \boxed{1} \end{bmatrix} \\ J \begin{bmatrix} H\ \boxed{1} \end{bmatrix} \end{bmatrix}$$

$$\begin{bmatrix} F \begin{bmatrix} K\ a \end{bmatrix} \end{bmatrix} \quad \begin{bmatrix} J \begin{bmatrix} K\ a \end{bmatrix} \end{bmatrix} \quad \begin{bmatrix} J \begin{bmatrix} H\ \boxed{1} \end{bmatrix} \\ G\ \boxed{1} \end{bmatrix} \quad \begin{bmatrix} F \begin{bmatrix} H\ \boxed{1} \end{bmatrix} \\ J \begin{bmatrix} K\ \boxed{1} \end{bmatrix} \end{bmatrix}$$

$$\begin{bmatrix} F \begin{bmatrix} K\ *top* \end{bmatrix} \end{bmatrix} \quad \begin{bmatrix} J \begin{bmatrix} K\ *top* \end{bmatrix} \end{bmatrix} \quad \begin{bmatrix} J \begin{bmatrix} K\ \boxed{1} \end{bmatrix} \\ G\ \boxed{1} \end{bmatrix} \quad \begin{bmatrix} J \begin{bmatrix} H\ \boxed{1} \\ K\ \boxed{1} \end{bmatrix} \end{bmatrix}$$

$$\begin{bmatrix} F \begin{bmatrix} H\ a \end{bmatrix} \end{bmatrix} \quad \begin{bmatrix} J \begin{bmatrix} H\ a \end{bmatrix} \end{bmatrix}$$

$$\begin{bmatrix} F \begin{bmatrix} H\ *top* \end{bmatrix} \end{bmatrix} \quad \begin{bmatrix} J \begin{bmatrix} H\ *top* \end{bmatrix} \end{bmatrix} \quad \begin{bmatrix} F \begin{bmatrix} H\ \boxed{1} \\ K\ \boxed{1} \end{bmatrix} \end{bmatrix}$$

When we come to talk about unification in the next section, it will be useful to think in terms of the individual pieces of information that make up a TFS.

3.4 Unification

We have seen some examples of unification already, but now I will discuss it in more detail. Unification is the combination of two TFSs to give the most general TFS which retains all the information which they individually contain. If there is no such TFS, unification is said to fail.

To make this description precise, we must first discuss the idea of generality with respect to TFSs. TFSs can be regarded as being ordered by specificity. Unlike the type hierarchy, where specificity is stipulated by the grammar writer, TFS specificity can be determined automatically, based on a notion of the information the TFSs contain. For instance, the TFS in the last exercise, repeated here as 3.21, contains more information than the structure in 3.22.

$$(3.21) \quad \begin{bmatrix} t \\ \text{F } \boxed{0} \begin{bmatrix} u \\ \text{H } \boxed{1}\,a \\ \text{K } \boxed{1} \end{bmatrix} \\ \text{G } \boxed{1} \\ \text{J } \boxed{0} \end{bmatrix}$$

$$(3.22) \quad \begin{bmatrix} t \\ \text{F } \boxed{0} \begin{bmatrix} u \\ \text{H } \boxed{1}\,a \\ \text{K } \boxed{1} \end{bmatrix} \\ \text{G } a \\ \text{J } \boxed{0} \end{bmatrix}$$

3.21 specifies that G and F.H are equivalent (and also G and F.K, G and J.H, G and J.K). 3.22 leaves this open and contains no information that isn't in 3.21. Thus 3.21 is strictly more general than 3.22. The technical term for this is that the more general structure *subsumes* the less general one. Consider the path-value and path-path equivalences discussed in the last exercise. If you construct the equivalences for 3.22, you will find they are a strict subset of the ones for 3.21.

The more specific structure will always have all the paths and path equivalences of the more general structure, and may have additional paths and path equivalences. The subsumption relationship is also controlled by the types on the paths. The more general structure must have types for its paths that are either equal to or more general than those for the corresponding paths in the more specific structure. For instance, 3.23 is more general than (subsumes) 3.24. If we assume that **c** is a subtype of **b**, then 3.24 in turn subsumes 3.25.

$$(3.23) \quad \begin{bmatrix} t \\ \text{G } a \\ \text{J *top*} \end{bmatrix}$$

$$(3.24) \quad \begin{bmatrix} t \\ \text{G } a \\ \text{J } b \end{bmatrix}$$

$$(3.25) \quad \begin{bmatrix} t \\ \text{G } a \\ \text{J } c \end{bmatrix}$$

(This illustrates why the exercise above asked you to construct the pairings of the path and the node's supertypes, as well as its actual type: this means subsumption corresponds to a subset relationship between the sets of decomposed elements of a TFS.)

The most general TFS of all is always $\left[\, \textbf{*top*} \,\right]$.

Subsumption can now be described more formally and concisely.

Properties of subsumption

> A TFS FS1 subsumes another TFS FS2 if and only if the following conditions hold:
>
> **Path values** For every path P in FS1 with a value of type **t**, there is a corresponding path P in FS2 with a value which is either **t** or a subtype of **t**.
>
> **Path equivalences** Every pair of paths P and Q which are reentrant in FS1 (i.e., which lead to the same node in the graph) are also reentrant in FS2.

Unification can now be defined very concisely in terms of subsumption.

Properties of unification

> The unification of two TFSs FS1 and FS2 is the most general TFS which is subsumed by both FS1 and FS2, if it exists.

It follows from this definition that if one of the structures specifies that a node at the end of some path P has a type **a**, and in the other structure path P leads to a node of type **b**, the structures will only unify if **a** and **b** are compatible types. If they are compatible, the node in the result will have the type which is the greatest lower bound of **a** and **b**. Thus unification of TFSs is always defined with respect to a particular type hierarchy. Another way of putting this definition is that if we take a hierarchy of TFSs ordered by subsumption, the result of unification corresponds to the greatest lower bound of the structures being unified.

An alternative way of looking at unification, in terms of the decomposition of structures, is that unification corresponds to taking the union of the sets of path-type and path-path equivalences from each of the structures to be unified and trying to form the resulting set of structures into a single TFS (see exercise 2 at the end of this section).

The symbol I will use for unification is $\sqcap$.[21] For instance, assuming

---

[21]Some other authors, including Sag and Wasow (1999), use $\sqcup$. The reason for the difference in directionality basically dates back to authors who used alternative ways of formalising unification, some of which roughly correspond to the alternative viewpoints I just mentioned. The LKB documentation has always used $\sqcap$ simply because it's more consistent with drawing type hierarchies with the most general type topmost, which most people seem to find the most natural direction.

the type hierarchy from 3.7:

$$(3.26) \quad \begin{bmatrix} \textbf{word} \\ \text{ORTH "these"} \end{bmatrix} \sqcap \begin{bmatrix} \textbf{word} \\ \text{CATEG } \textbf{np} \end{bmatrix} = \begin{bmatrix} \textbf{word} \\ \text{ORTH "these"} \\ \text{CATEG } \textbf{np} \end{bmatrix}$$

The root nodes of the two structures being unified always correspond to the root node of the result but arcs with different features always give distinct arcs in the result. Contrast this example with:

$$(3.27) \quad \begin{bmatrix} \textbf{word} \\ \text{ORTH "these"} \\ \text{CATEG } \textbf{*top*} \end{bmatrix} \sqcap \begin{bmatrix} \textbf{word} \\ \text{CATEG } \textbf{np} \end{bmatrix} = \begin{bmatrix} \textbf{word} \\ \text{ORTH "these"} \\ \text{CATEG } \textbf{np} \end{bmatrix}$$

Here both the TFSs being unified have an arc labelled CATEG and this must give a single arc in the result (since TFSs may only have one arc with a given feature from any node). The first structure says that the value of CATEG is ***top***, the second that it is **np**, but since these types are consistent, the type on the node in the result is simply their greatest lower bound: that is **np**.

Unification, like other mathematical operations, can be regarded procedurally or statically. For instance, one can equivalently say 'the result of adding 2 and 3 is 5' which suggests a procedure, or '5 is the sum of 2 plus 3', or $5 = 2 + 3$. With respect to unification, terms like 'failure' are somewhat procedural. Because of this, it is useful to introduce a symbol that stands for inconsistency, $\perp$ (bottom). For instance, the unification of the following two structures is $\perp$.

$$(3.28) \quad \begin{bmatrix} \textbf{syn-struc} \\ \text{CATEG } \textbf{vp} \end{bmatrix} \sqcap \begin{bmatrix} \textbf{syn-struc} \\ \text{CATEG } \textbf{np} \end{bmatrix} = \perp$$

The inconsistency arises because of the inconsistent types for the path CATEG (i.e., **np** and **vp** don't have a glb).

I will now go through some further examples to illustrate unification in more detail.

3.4.1 Examples of unification

Example 1

$$\begin{bmatrix} \textbf{syn-struc} \\ \text{CATEG } \textbf{np} \\ \text{ARGS} \begin{bmatrix} \textbf{*ne-list*} \\ \text{FIRST} \begin{bmatrix} \textbf{syn-struc} \\ \text{CATEG } \textbf{*top*} \end{bmatrix} \end{bmatrix} \end{bmatrix} \sqcap \begin{bmatrix} \textbf{phrase} \\ \text{ARGS} \begin{bmatrix} \textbf{*ne-list*} \\ \text{FIRST} \begin{bmatrix} \textbf{phrase} \\ \text{CATEG } \textbf{vp} \end{bmatrix} \\ \text{REST } \textbf{word} \end{bmatrix} \end{bmatrix}$$

$$= \begin{bmatrix} \textbf{phrase} \\ \text{CATEG } \textbf{np} \\ \text{ARGS} \begin{bmatrix} \textbf{*ne-list*} \\ \text{FIRST} \begin{bmatrix} \textbf{phrase} \\ \text{CATEG } \textbf{vp} \end{bmatrix} \\ \text{REST } \textbf{word} \end{bmatrix} \end{bmatrix}$$

In this example, we have to consider paths of length greater than one, but unification of the substructures works in exactly the same way. For instance, the root node of the result has the type **phrase** because **phrase** is the glb of **phrase** and **syn-struc**. Similarly, the node at the end of the path ARGS.FIRST also has the type **phrase**.

Example 2

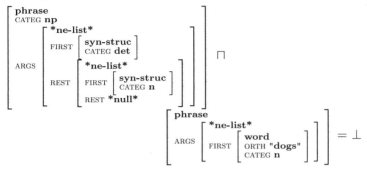

Here the value of the path ARGS.FIRST.CATEGORY is **det** in the first structure and **n** in the second. These types are not compatible, so unification fails.

Example 3

The point about this example is to illustrate that if one of the input structures subsumes the other, the result is the most specific structure. As I mentioned above, the most general TFS of all is [***top***]: the result of unifying this with an arbitrary TFS F will always be F. Note also that if two identical TFSs G are unified, the result will be G.

Example 4

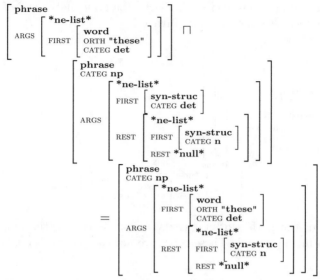

The result in this example is the TFS I showed in 3.18. This is a partial structure which might be created in the course of parsing with the grammar in Figure 5, because it illustrates a partial application of the NP rule (the second argument to the unification) to the lexical entry for *these*. Because the structure corresponding to the lexical entry has to go in the ARGS.FIRST position, the first argument to the unification is a sort of skeleton structure with the lexical structure at the end of the ARGS.FIRST path.

Example 5

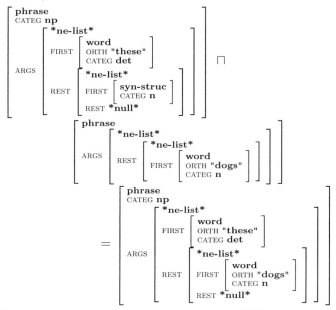

Here the result from example 4 is the first TFS: it is unified with a TFS representing a second daughter for the rule application, corresponding to the lexical entry for *dogs* (this time the actual lexical entry structure is at the end of the ARGS.REST.FIRST path, since this corresponds to the second daughter position). The result is a structure which is a complete representation of the phrase *these dogs*. At this point I should emphasize a crucial property of unification, which is that the result is independent of the order in which we combine the structures (more formally, unification is commutative and transitive). For instance, if we take the daughter structures first, unify them together, and then unify them with the rule structure, the result will be identical to that above.

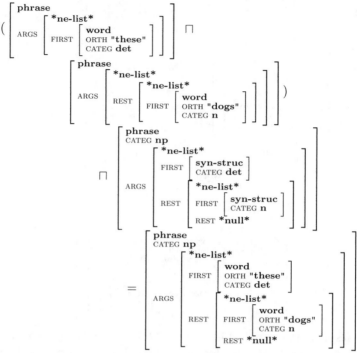

The effect of this is that we can guarantee that different parsing strategies will always give the same result with a typed feature structure grammar (if all parses are generated and the parsing process terminates and does not run into an infinite loop). However some strategies may be more efficient than others, since if unifying a set of structures results in failure, the process of determining this is quicker if the failure occurs as one of the early unification steps. More details of parsing are given in Chapter 4.

Example 6

$$
\begin{bmatrix} t \\ \text{F} \ \boxed{1}\,\text{*top*} \\ \text{G} \ \boxed{1} \end{bmatrix}
\ \sqcap\
\begin{bmatrix} t \\ \text{F} \begin{bmatrix} u \\ \text{J a} \end{bmatrix} \\ \text{G} \begin{bmatrix} u \\ \text{J *top*} \\ \text{K b} \end{bmatrix} \end{bmatrix}
\ =\
\begin{bmatrix} t \\ \text{F} \ \boxed{0} \begin{bmatrix} u \\ \text{J a} \\ \text{K b} \end{bmatrix} \\ \text{G} \ \boxed{0} \end{bmatrix}
$$

This example involves reentrancy. I am using abstract TFSs for simplicity and assuming a flat type hierarchy, where **t**, **u**, **a** and **b** are all mutually incompatible daughters of ***top***. Note that the information on the reentrant node in the result all comes from the second structure: effectively the result of unification has been to combine two nodes which were distinct in the input structure.

Example 7

$$
\begin{bmatrix} \text{t} \\ \text{F}\ \boxed{1}\ \text{*top*} \\ \text{G}\ \boxed{1} \end{bmatrix}
\ \sqcap\
\begin{bmatrix} \text{t} \\ \text{F}\ \begin{bmatrix} \text{u} \\ \text{J}\ \text{a} \end{bmatrix} \\ \text{G}\ \begin{bmatrix} \text{u} \\ \text{J}\ \text{b} \\ \text{K}\ \text{b} \end{bmatrix} \end{bmatrix}
\ =\ \bot
$$

Here I modified the example above slightly, so that the value on G.J is **b** instead of **a**. Since **a** and **b** are incompatible, unification fails.

Example 8

$$
\begin{bmatrix} \text{t} \\ \text{F}\ \begin{bmatrix} \text{u} \\ \text{G}\ \boxed{1} \end{bmatrix} \\ \text{H}\ \boxed{1} \end{bmatrix}
\ \sqcap\
\begin{bmatrix} \text{t} \\ \text{F}\ \boxed{1} \\ \text{H}\ \boxed{1} \end{bmatrix}
\ =\ \bot
$$

This unification results in failure because the result would be a cyclic structure. Note that cyclic structures are the only cases where unification failure can occur without incompatible types being present.

3.4.2 *Formal definitions of subsumption and unification

I will use $\mathcal{F}$ to denotes the collection of TFSs. I use the notation $\pi \equiv_F \pi'$ to mean that TFS F contains path equivalence or reentrancy between the paths π and π' (i.e., $\delta(r, \pi) = \delta(r, \pi')$ where r is the root node of F); and $\mathcal{P}_F(\pi) = \sigma$ means that the type on the path π in F is σ (i.e., $\mathcal{P}_F(\pi) = \sigma$ if and only if $\theta(\delta(r, \pi)) = \sigma$, where r is the root node of F). Subsumption is then defined as follows:

F subsumes F', written $F' \sqsubseteq F$, if and only if:

- $\pi \equiv_F \pi'$ implies $\pi \equiv_{F'} \pi'$
- $\mathcal{P}_F(\pi) = t$ implies $\mathcal{P}_{F'}(\pi) = t'$ and $t' \sqsubseteq t$

The unification $F \sqcap F'$ of two TFSs F and F' is the greatest lower bound of F and F' in the collection of TFSs ordered by subsumption.

3.4.3 Exercises

These exercises involve more examples of unification.[22]

1. Give the results of the following unifications, assuming that the types **t**, **u**, **a** and **b** are incompatible daughters of ***top***.

---

[22]You could try using the LKB system to investigate unification, by specifying entries, viewing the TFSs, and unifying the results according to the instructions in §6.3.1. Be warned, however, that if you do this you will probably find that some of the entries you specify get expanded with additional information because of the well-formedness conditions. I will explain how this happens in detail in the next section.

(a) $\begin{bmatrix} \mathbf{t} \\ \text{F } \boxed{1}\ \mathbf{a} \\ \text{G } \boxed{1} \end{bmatrix}$ $\sqcap$ $\begin{bmatrix} \mathbf{t} \\ \text{G } \mathbf{b} \end{bmatrix}$

(b) $\begin{bmatrix} \mathbf{t} \\ \text{F } \boxed{1} \\ \text{G } \boxed{1} \\ \text{H } \boxed{2} \\ \text{J } \boxed{2} \end{bmatrix}$ $\sqcap$ $\begin{bmatrix} \mathbf{t} \\ \text{F } \boxed{1} \\ \text{J } \boxed{1} \end{bmatrix}$

(c) $\begin{bmatrix} \mathbf{t} \\ \text{F } \begin{bmatrix} \mathbf{u} \\ \text{G } \boxed{1} \end{bmatrix} \\ \text{H } \boxed{1} \end{bmatrix}$ $\sqcap$ $\begin{bmatrix} \mathbf{t} \\ \text{F } \boxed{2} \\ \text{H } \begin{bmatrix} \mathbf{u} \\ \text{J } \boxed{2} \end{bmatrix} \end{bmatrix}$

2. Consider the following example involving reentrancy:

$$\begin{bmatrix} \mathbf{t} \\ \text{F } \boxed{1}\mathbf{*top*} \\ \text{G } \boxed{1} \end{bmatrix} \quad \sqcap \quad \begin{bmatrix} \mathbf{t} \\ \text{F } \boxed{2}\mathbf{a} \\ \text{H } \boxed{2} \end{bmatrix} \quad = \quad \begin{bmatrix} \mathbf{t} \\ \text{F } \boxed{3}\mathbf{a} \\ \text{G } \boxed{3} \\ \text{H } \boxed{3} \end{bmatrix}$$

Write down the path-path and the path-value equivalences for each structure (assume the types **t** and **a** are immediate subtypes of ***top***). What can you say about the relationship between the set of equivalences for the result compared with those for the arguments to unification?

3. Assume the type hierarchy given for the grammar with agreement shown in Figure 7.

(a) What is the result of the following:

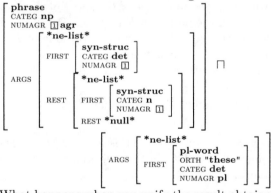

(b) What happens when you unify the result obtained above with the following structure?

$$\begin{bmatrix} \text{ARGS} \begin{bmatrix} \mathbf{*ne\text{-}list*} \\ \text{REST} \begin{bmatrix} \mathbf{*ne\text{-}list*} \\ \text{FIRST} \begin{bmatrix} \mathbf{pl\text{-}word} \\ \text{ORTH } \mathbf{"dogs"} \\ \text{CATEG } \mathbf{n} \\ \text{NUMAGR } \mathbf{pl} \end{bmatrix} \end{bmatrix} \end{bmatrix} \end{bmatrix}$$

(c) What about this one?

$$
\begin{bmatrix}
\text{ARGS} & \begin{bmatrix}
\textbf{*ne-list*} \\
\text{REST} & \begin{bmatrix}
\textbf{*ne-list*} \\
\text{FIRST} & \begin{bmatrix}
\textbf{word} \\
\text{ORTH "dogs"} \\
\text{CATEG } \textbf{n} \\
\text{NUMAGR } \textbf{agr}
\end{bmatrix}
\end{bmatrix}
\end{bmatrix}
\end{bmatrix}
$$

(d) And this?

$$
\begin{bmatrix}
\text{ARGS} & \begin{bmatrix}
\textbf{*ne-list*} \\
\text{REST} & \begin{bmatrix}
\textbf{*ne-list*} \\
\text{FIRST} & \begin{bmatrix}
\textbf{sg-word} \\
\text{ORTH "dog"} \\
\text{CATEG } \textbf{n} \\
\text{NUMAGR } \textbf{sg}
\end{bmatrix}
\end{bmatrix}
\end{bmatrix}
\end{bmatrix}
$$

4. Assume that **t** is an immediate daughter of ***top***. Consider the following structure:

$$
\begin{bmatrix}
\textbf{t} \\
\text{F} \ \boxed{1}\textbf{*top*} \\
\text{G} \ \boxed{1}
\end{bmatrix}
$$

How many structures are there which subsume it? Draw these structures in a subsumption hierarchy, with the most general structure at the top, the intermediate structures arranged in order of generality and the full structure at the bottom.

5. (Optional) Suppose we are told that

$$
FS1 \sqcap FS2 = \begin{bmatrix}
\textbf{t} \\
\text{F} \ \boxed{1}\textbf{*top*} \\
\text{G} \ \boxed{1}
\end{bmatrix}
$$

but we don't know the structures FS1 and FS2. How many different pairs of TFSs FS1 and FS2 satisfy the equation? Why might this sort of question be relevant to language processing?

3.4.4 Answers

1. (a) ⊥

   (b)
   $$
   \begin{bmatrix}
   \textbf{t} \\
   \text{F} \ \boxed{1} \\
   \text{G} \ \boxed{1} \\
   \text{H} \ \boxed{1} \\
   \text{J} \ \boxed{1}
   \end{bmatrix}
   $$

   (c) ⊥ (cyclic structure).

2. (a)

path	type		path	path
empty path	**t**		F	G
empty path	***top***			
F	***top***			
G	***top***			

(b)

path	type
empty path	**t**
empty path	***top***
F	***top***
F	**a**
H	***top***
H	**a**

path	path
F	H

(c)

path	type
empty path	**t**
empty path	***top***
F	***top***
F	**a**
G	***top***
G	**a**
H	***top***
H	**a**

path	path
F	G
F	H
G	H

The result contains all the equivalences from the arguments plus one extra path-value equivalence between G and **a** and one extra path-path equivalence between G and H. The extra pieces arise because of the reentrancy. In general, the path-value plus path-path equivalences for the TFS which is the result of a successful unification are a superset of the union of those of the arguments.

3. (a)

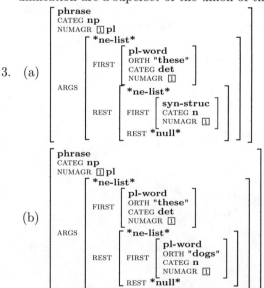

(b)

(c)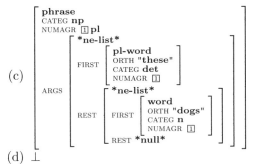

(d) ⊥

As you have probably realized, this demonstrates how reentrancy ensures that the daughters of the np rule have consistent values for agreement. Note that when unification succeeds, the resulting phrase has a value for NUMAGR which is the unification of the values of NUMAGR on the daughters. This ensures that there is also agreement with the verb phrase. With respect to the last example, notice that the three TFSs which were unified were all pairwise compatible: it was only the combination of all three which fails. Thus the np rule itself works equally well for both singular and plural cases.

4. On the assumption that **t** is an immediate daughter of ***top***, there are 10 TFSs that subsume

$$\begin{bmatrix} \mathbf{t} \\ \text{F } \boxed{1} \ \mathbf{*top*} \\ \text{G } \boxed{1} \end{bmatrix}$$

(including the structure itself). These are shown in Figure 8. If you examine this hierarchy, you will see that any two TFSs always have a unique greatest lower bound, which is equal to their unification. You should also be able to see that every pair of structures has a unique least upper bound (lub). This corresponds to finding a TFS which represents the pieces of information that common to both structures. This operation is called *generalisation* — it is currently relevant in the LKB only with respect to the treatment of defaults, which we are not going to discuss in this book.

5. Answer to optional part: There are 14 possible pairs of structures (or 27 if you count ordered pairs) which can be combined to give this structure. I won't list them all here, but briefly go through the argument. We know the only structures that can be involved are those that subsume the full structure. We can consider each of these 10 possibilities for FS1 and see how many possible structures for FS2 there are in each case, ignoring any that duplicate those we have already found. If FS1 is the full structure, there are 10

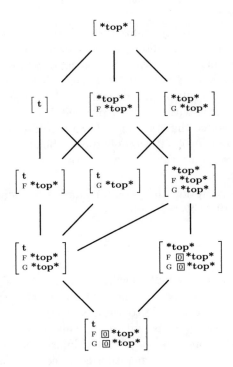

FIGURE 8 Subsumption hierarchy

possibilities for FS2. If FS1 is:

$$\begin{bmatrix} \textbf{*top*} \\ \text{F } \boxed{1}\textbf{*top*} \\ \text{G } \boxed{1} \end{bmatrix}$$

there are 5 possibilities for FS2 (all the structures which specify that the root node is of type **t**), but one of these is the case
where FS2 is the full structure, which is equivalent to a pair we
have already considered. Of the other 8 possibilities for FS1, none
contain any information about reentrancy, thus they must be combined with a reentrant structure, but we've already considered all
these cases.

In general, if $F \sqcap G = H$, then we know that both F and G subsume (or are equal to) H, and that they mutually contain all the
information in H, but we have no way of knowing which information is in which structure. There will be a finite number of possible
candidate TFS pairs (assuming H is a finite structure), but there
could be a very large number of possibilities. Even if we also know
F, we still cannot in general determine G, because while we know
that it must contain all the information that is in H which is not
in F, it may also contain some of the same information as F. The
only exception to this is if F is the most general structure $\begin{bmatrix} \textbf{*top*} \end{bmatrix}$,
then G must be equal to H.

The relevance of this concerns processing strategies. While we can
deterministically and efficiently unify two structures, we cannot,
in general, reverse the operation efficiently. The claim is often
made that unification-based grammars are *reversible* or *bidirectional*: what is meant by this is that they can be used for parsing
and generation. To be accurate this statement has to be qualified
in multiple ways — it certainly isn't true that all unification-based
grammars are suitable for generation. But the point here is just
that efficient grammar reversibility does not involve reversing unifications.

3.5   Type constraints and inheritance

We finally arrive at a detailed description of the operation of the type
constraints. In the previous sections, we have been discussing TFSs in
general, but now I want to concentrate on those which are well-formed
with respect to a set of type constraints. Usually the TFSs corresponding
directly to descriptions (e.g., of lexical entries) won't be well-formed:
the process of inheriting/inferring information, which I've alluded to
a couple of times informally so far, is precisely the process of making
the structure well-formed. The primary purpose of type constraints as

far as the grammar writer is concerned is that they can be used to allow generalizations to be expressed, so that lexical entries and other descriptions can be kept succinct. Their secondary purpose is to avoid errors creeping into a grammar, such as misspelt feature names.

In the LKB system, the constraint on a type is expressed as a TFS. I will start off by defining well-formedness of TFSs in general with respect to a set of type constraints, and talk about how the system converts a non-well-formed structure to a well-formed one, via type inference. Then we'll look at how unification operates on well-formed TFSs. Finally I will describe the internal conditions on a set of type constraints, and discuss how the definitions of types we have seen in the examples so far (e.g., in Figure 5) give rise to the actual type constraints.

3.5.1 Well-formedness

First let's say that the *substructures* of a TFS are the TFSs rooted at each node in the structure. Then we can describe well-formedness in terms of conditions on each substructure. We'll also talk about the features which label arcs starting from the root node of a structure as the *top-level* features of a structure. The top-level features of a type constraint are referred to as the *appropriate features* for that type.

Properties of a well-formed TFS

> **Constraint** Each substructure of a well-formed TFS must be subsumed by the constraint corresponding to the type on the substructure's root node.
>
> **Appropriate features** The top-level features for each substructure of a well-formed TFS must be the appropriate features of the type on the substructure's root node.

Figure 9 shows the full constraint and the appropriate features for all the types in the grammar **g1cfg** (i.e., the grammar shown in Figure 5). Note that the constraints on some types, such as **phrase**, contain some information which was not in their description: this is because they have inherited information from types higher in the hierarchy, as we'll see in detail in §3.5.8.

Some types, such as **cat**, **n** and **vp** in the example grammar, have no appropriate features. This means that in any well-formed TFS, they can only label terminal nodes. I will refer to types which have no appropriate features and which have no descendants with appropriate features as *atomic* types. The type ***top*** has no appropriate features itself, but some of its descendants do, so it is not an atomic type.

The following examples are all well-formed TFSs given these type constraints (note that not all well-formed structures are sensible!):

type	constraint	appropriate features
top	$\left[\, \textbf{*top*} \,\right]$	
string	$\left[\, \textbf{string} \,\right]$	
list	$\left[\, \textbf{*list*} \,\right]$	
ne-list	$\begin{bmatrix} \textbf{*ne-list*} \\ \text{FIRST } \textbf{*top*} \\ \text{REST } \textbf{*list*} \end{bmatrix}$	FIRST REST
null	$\left[\, \textbf{*null*} \,\right]$	
syn-struc	$\begin{bmatrix} \textbf{syn-struc} \\ \text{CATEG } \textbf{cat} \end{bmatrix}$	CATEG
cat	$\left[\, \textbf{cat} \,\right]$	
s	$\left[\, \textbf{s} \,\right]$	
np	$\left[\, \textbf{np} \,\right]$	
vp	$\left[\, \textbf{vp} \,\right]$	
det	$\left[\, \textbf{det} \,\right]$	
n	$\left[\, \textbf{n} \,\right]$	
phrase	$\begin{bmatrix} \textbf{phrase} \\ \text{CATEG } \textbf{cat} \\ \text{ARGS } \textbf{*list*} \end{bmatrix}$	CATEG ARGS
word	$\begin{bmatrix} \textbf{word} \\ \text{ORTH } \textbf{string} \\ \text{CATEG } \textbf{cat} \end{bmatrix}$	CATEG ORTH

FIGURE 9 Constraints and appropriate features for grammar `g1cfg`

(3.29)
$$\begin{bmatrix} \textbf{phrase} \\ \text{CATEG } \textbf{s} \\ \text{ARGS } \textbf{*list*} \end{bmatrix}$$

(3.30)
$$\begin{bmatrix} \textbf{word} \\ \text{ORTH } \textbf{"on"} \\ \text{CATEG } \textbf{s} \end{bmatrix}$$

(3.31)
$$\begin{bmatrix} \textbf{*ne-list*} \\ \text{FIRST } \boxed{1} \textbf{*list*} \\ \text{REST } \boxed{1} \end{bmatrix}$$

(3.32)
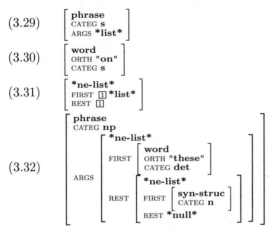

The following examples are not well-formed and do not subsume well-formed structures:

(3.33)
$$\begin{bmatrix} \textbf{phrase} \\ \text{CATEG } \textbf{word} \\ \text{ARGS } \textbf{*list*} \end{bmatrix}$$
Wrong type on CATEG.

(3.34)
$$\begin{bmatrix} \textbf{syn-struc} \\ \text{CATEG } \textbf{s} \\ \text{AGS } \textbf{*list*} \end{bmatrix}$$
AGS is not an appropriate feature for **syn-struc** (or for anything else).

(3.35)
$$\begin{bmatrix} \textbf{syn-struc} \\ \text{CATEG } \textbf{s} \\ \text{FIRST } \textbf{word} \end{bmatrix}$$
FIRST is not an appropriate feature for **syn-struc**.

The following examples are not well-formed but they subsume well-formed structures which can be constructed by the type inference process which I will discuss below:

(3.36)
$$\begin{bmatrix} \textbf{phrase} \\ \text{CATEG } \textbf{*top*} \\ \text{ARGS } \textbf{*list*} \end{bmatrix}$$
Wrong type on CATEG, but it is compatible with a valid type.

(3.37)
$$\begin{bmatrix} \textbf{syn-struc} \\ \text{CATEG } \textbf{s} \\ \text{ARGS } \textbf{*list*} \end{bmatrix}$$
ARGS is not an appropriate feature for **syn-struc** but it is appropriate for **phrase** which is a subtype of **syn-struc**.

(3.38)
$$\begin{bmatrix} \textbf{phrase} \\ \text{CATEG } \textbf{s} \end{bmatrix}$$
The ARGS feature is missing, so the constraint on **phrase** does not subsume this structure, but it is compatible with it.

3.5.2 *Formal definition of well-formedness

The constraint function is given by $C: \langle \mathsf{Type}, \sqsubseteq \rangle \to \mathcal{F}$.

For each type there is a set of features appropriate to that type: $Appfeat: \langle \mathsf{Type}, \sqsubseteq \rangle \to \mathsf{Feat}$. The appropriate features for a type are determined by its constraint:

If $C(t) = \langle Q, q_0, \delta, \theta \rangle$ then the appropriate features of t are defined as $Appfeat(t) = Feat(q_0)$ where $Feat(q)$ is defined to be the set of features labeling transitions from the node q — i.e., $f \in Feat(q)$ such that $\delta(q, f)$ is defined.

We say that a given TFS $F = \langle Q, q_0, \delta, \theta \rangle$ is a well-formed TFS iff for all nodes $q \in Q$, the substructure TFS rooted at q, F', is such that $F' = \langle Q', q, \delta', \theta' \rangle \sqsubseteq C(\theta(q))$ and $Feat(q) = Appfeat(\theta(q))$.

3.5.3 Exercises on well-formedness

1. For each of the following, specify whether it is well-formed or not (assuming the constraints given in Figure 9). For the non-well-formed structures which subsume some well-formed structures, list the most general such well-formed structure.

   (a) $\begin{bmatrix} \textbf{*null*} \\ \text{FIRST } \textbf{syn-struc} \end{bmatrix}$

   (b) $\begin{bmatrix} \textbf{*top*} \\ \text{FIRST } \textbf{syn-struc} \end{bmatrix}$

   (c) $\begin{bmatrix} \textbf{*top*} \\ \text{REST } \left[\text{REST } \textbf{*null*} \right] \end{bmatrix}$

   (d) $\begin{bmatrix} \textbf{*top*} \\ \text{CATEG } \boxed{0} \\ \text{ARGS } \left[\text{REST } \boxed{0} \begin{bmatrix} \textbf{*top*} \\ \text{REST } \textbf{*top*} \end{bmatrix} \right] \end{bmatrix}$

2. Consider the subsumption hierarchy you constructed in the answers to the exercises in §3.4.3. Which of the structures are well-formed feature structures? Assume that the type **t** is an immediate daughter of ***top***, and has the following constraint:

   $\begin{bmatrix} \text{t} \\ \text{F } \textbf{*top*} \\ \text{G } \textbf{*top*} \end{bmatrix}$

3.5.4 Answers to exercises on well-formedness

1. (a) not well formed — FIRST isn't an appropriate feature for ***null***.

   (b) not well formed, but subsumes:

   $\begin{bmatrix} \textbf{*ne-list*} \\ \text{FIRST } \begin{bmatrix} \textbf{syn-struc} \\ \text{CATEG } \textbf{cat} \end{bmatrix} \\ \text{REST } \textbf{*list*} \end{bmatrix}$

(c) not well formed, but subsumes:

$$\begin{bmatrix} \textbf{*ne-list*} \\ \text{FIRST } \textbf{*top*} \\ \text{REST } \begin{bmatrix} \textbf{*ne-list*} \\ \text{FIRST } \textbf{*top*} \\ \text{REST } \textbf{*null*} \end{bmatrix} \end{bmatrix}$$

(d) not well formed — CATEG's value must be of type **cat** but this node is reentrant with a node that must be of type ***ne-list***, which isn't compatible with **cat**.

2. The only well-formed structures are the following (shown in a subsumption hierarchy):

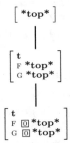

In general, we can talk about a subsumption hierarchy of well-formed TFSs, and define well-formed unification as determining the greatest lower bound within that hierarchy.

3.5.5 Type inference

Type inference takes a non-well-formed TFS and returns the most general well-formed structure which it subsumes. Thus it always preserves the information in the initial structure. It is always possible to find a unique most general well-formed structure from a non-well-formed structure if the latter subsumes any well-formed structures. The LKB system carries out type inference on all entries (i.e., lexical entries, grammar rules, and so on). It is an error to define an entry which cannot be converted into a well-formed TFS.

The first part of type inference consists of specialising the types on each node in the structure to the most general type for which all the node's features are appropriate. For instance, in the following structure, ARGS is not an appropriate feature for **syn-struc** but it is appropriate for **phrase**, which is a subtype of **syn-struc**.

(3.39) $\begin{bmatrix} \textbf{syn-struc} \\ \text{CATEG } \textbf{s} \\ \text{ARGS } \textbf{*top*} \end{bmatrix}$

The result is therefore

(3.40) $\begin{bmatrix} \textbf{phrase} \\ \text{CATEG } \textbf{s} \\ \text{ARGS } \textbf{*top*} \end{bmatrix}$

If the initial structure had been:

$$(3.41) \quad \begin{bmatrix} \text{*top*} \\ \text{CATEG } \mathbf{s} \\ \text{FIRST } \mathbf{word} \end{bmatrix}$$

we could not have found a possible type, because there are no types for which both CATEG and FIRST are appropriate. If it had been:

$$(3.42) \quad \begin{bmatrix} \text{*list*} \\ \text{CATEG } \mathbf{s} \end{bmatrix}$$

we also could not have found a possible type, because although CATEG is an appropriate feature for **syn-struc**, this is not a subtype of ***list***. If the initial structure had been:

$$(3.43) \quad \begin{bmatrix} \text{*top*} \\ \text{CATEG } \mathbf{s} \end{bmatrix}$$

the type found would have been **syn-struc** rather than **phrase** or **word** because type inference always returns the most general possible type. There is a condition on how features are introduced in the type constraints that guarantees there will always be a unique most general type for any set of features if that set of features is appropriate for any types. I will discuss this in §3.5.8.

The second stage in type inference consists of ensuring that all nodes are subsumed by their respective type constraints. This involves unifying each node with the constraint of the type on each node (using well-formed unification, described below in §3.5.6). So from

$$(3.44) \quad \begin{bmatrix} \text{phrase} \\ \text{CATEG } \mathbf{s} \\ \text{ARGS *top*} \end{bmatrix}$$

we obtain

$$(3.45) \quad \begin{bmatrix} \text{phrase} \\ \text{CATEG } \mathbf{s} \\ \text{ARGS *list*} \end{bmatrix}$$

Type inference may also fail at this point. For instance, given the structure:

$$(3.46) \quad \begin{bmatrix} \text{phrase} \\ \text{CATEG *list*} \\ \text{ARGS *top*} \end{bmatrix}$$

we find that the type on CATEG ***list*** is inconsistent with the constraint on **phrase**, which stipulates that the value of CATEG must be the type **cat**.

3.5.6 Unification revisited

We want unification of well-formed TFSs to give a well-formed result. We can achieve this by giving a definition of the unification of well-formed TFSs which parallels unification of TFSs in general.

Properties of well-formed unification

The well-formed unification of two TFSs F and G is the most general well-formed TFS which is subsumed by both F and G, if it exists.

This makes sense because we can talk about a subsumption hierarchy of well-formed TFSs which is a subpart of the hierarchy of TFSs in general (see exercise 2, above). In most cases, this gives exactly the same result as we've seen from unification previously. However, there is a complication which arises only in some cases where there is multiple inheritance, specifically when there is a constraint on a type which is more specific than the combined constraints of two ancestor types.

To illustrate this, consider the corrected animal hierarchy shown in Figure 3.14, and suppose it also contains the type **boolean**, as a daughter of ***top***, and **true** and **false** inheriting from **boolean**. Suppose that the constraint on the type **swimmer** is 3.47, the constraint on **mammal** is 3.48, and the constraint on **whale** is 3.49:

$$(3.47) \quad \begin{bmatrix} \textbf{swimmer} \\ \text{FINS } \textbf{boolean} \end{bmatrix}$$

$$(3.48) \quad \begin{bmatrix} \textbf{mammal} \\ \text{FRIENDLY } \textbf{boolean} \end{bmatrix}$$

$$(3.49) \quad \begin{bmatrix} \textbf{whale} \\ \text{HARPOONED } \textbf{boolean} \\ \text{FINS } \textbf{true} \\ \text{FRIENDLY } \textbf{boolean} \end{bmatrix}$$

Consider what happens when we unify the following TFSs (both of which are well-formed):

$$(3.50) \quad \begin{bmatrix} \textbf{mammal} \\ \text{FRIENDLY } \textbf{true} \end{bmatrix} \sqcap \begin{bmatrix} \textbf{swimmer} \\ \text{FINS } \textbf{boolean} \end{bmatrix}$$

If we ignored the requirement for well-formedness, the result would be:

$$(3.51) \quad \begin{bmatrix} \textbf{whale} \\ \text{FINS } \textbf{boolean} \\ \text{FRIENDLY } \textbf{true} \end{bmatrix}$$

but this isn't well-formed — it lacks the feature HARPOONED, and the value of FINS isn't **true**. To get a well-formed feature structure, we also have to add the constraint information on **whale**, to get:

$$(3.52) \quad \begin{bmatrix} \textbf{whale} \\ \text{HARPOONED } \textbf{boolean} \\ \text{FINS } \textbf{true} \\ \text{FRIENDLY } \textbf{true} \end{bmatrix}$$

This result is the most general well-formed structure that contains all the information in the structures being unified, but of course it also contains additional information, derived from the type system.

If the constraint on the type that is the greatest lower bound is inconsistent with the given information, unification fails. For instance, if $\sqcap_{\text{wf}}$ refers to the operation of well-formed unification:

$$\begin{bmatrix} \textbf{mammal} \\ \text{FRIENDLY } \textbf{true} \end{bmatrix} \sqcap_{\text{wf}} \begin{bmatrix} \textbf{swimmer} \\ \text{FINS } \textbf{false} \end{bmatrix} = \perp$$

From now on, when I talk about unification of well-formed structures, I'll always be referring to this operation and I will use $\sqcap$ to refer to it.

3.5.7 *Formal definition of well-formed unification

The well-formed unification $F \sqcap_{\text{wf}} F'$ of two TFSs F and F' is the greatest lower bound of F and F' in the collection of well-formed TFSs ordered by subsumption.

3.5.8 Conditions on type constraints

The final part of the description of the typed feature structure formalism concerns the construction of the full type constraints from the descriptions and the conditions on the type constraints. This will show precisely how we get from the descriptions shown in Figure 5 to the constraints shown in Figure 9. I will refer to the feature structures which are directly specified in the descriptions as the *local constraints*. There are a series of conditions on full type constraints which determine how the local constraints are expanded into the full constraints.

Properties of type constraints

Type The type of the TFS expressing the constraint on a type **t** is always **t**.

Consistent inheritance The constraint on a type must be subsumed by the constraints on all its parents. This means that any local constraint specification must be compatible with the inherited information, and that in the case of multiple inheritance, the parents' constraints must unify.

Maximal introduction of features Any feature must be introduced at a single point in the hierarchy. That is, if a feature, F, is an appropriate feature for some type, **t**, and not an appropriate feature for any of its ancestors, then F cannot be appropriate for a type which is not a descendant of **t**. Note that the consistent inheritance condition guarantees that the feature will be appropriate for all descendants of **t**.

Well-formedness of constraints All full constraint feature structures must be well-formed as described in §3.5.1.

You should check Figure 9 and Figure 5 to make sure you follow how this applies to grammar `g1cfg`.

Although the constraints in the grammars we've been looking at are all very simple, constraints can in general be arbitrarily complex TFSs.

For instance, taking the version of the grammar with agreement, shown in Figure 7, we could have simplified the description of the rules that was given there by making them a TFS of type **agr-phrase**, where **agr-phrase** is a subtype of **phrase** with the following description:

(3.53) agr-phrase := phrase &
        [NUMAGR #1,
          ARGS [FIRST [NUMAGR #1],
                 REST [FIRST [NUMAGR #1],
                        REST *null*]]] .

There is one non-obvious consequence of the conditions on type constraints, which is that they disallow type descriptions such as the following:

(3.54) list := *top* &
        [FIRST *top*,
          REST list].

The reason is that this description would have to result in an infinite constraint structure when we tried to make it well-formed. That is, it would be expanded so that the TFS which is the value of REST would have features FIRST and REST and that value of the REST feature would be of type **list** which would be expanded in the same way, and so on, as indicated below:

$$(3.55) \quad \begin{bmatrix} \textbf{*ne-list*} \\ \text{FIRST } \textbf{*top*} \\ \text{REST} \begin{bmatrix} \textbf{*ne-list*} \\ \text{FIRST } \textbf{*top*} \\ \text{REST} \begin{bmatrix} \textbf{*ne-list*} \\ \text{FIRST } \textbf{*top*} \\ \text{REST} \dots \end{bmatrix} \end{bmatrix} \end{bmatrix}$$

The solution to this problem is to define lists in the way we've done in the example grammars, so that there are distinct subtypes for empty and non-empty lists, with the latter having no appropriate features.

(3.56)

        list := *top*.   *ne-list* := *list* &
                                  [FIRST *top*,
        null := *list*.   REST *list*].

With this definition, the previously impossible structure can be typed without causing an infinite structure:

$$(3.57) \quad \begin{bmatrix} \textbf{*ne-list*} \\ \text{FIRST } \textbf{*top*} \\ \text{REST } \textbf{*ne-list*} \end{bmatrix}$$

is expanded to:

(3.58)
$$\begin{bmatrix} \textbf{*ne-list*} \\ \text{FIRST } \textbf{*top*} \\ \text{REST} \begin{bmatrix} \textbf{*ne-list*} \\ \text{FIRST } \textbf{*top*} \\ \text{REST } \textbf{*list*} \end{bmatrix} \end{bmatrix}$$

3.5.9 *Formal definition of the constraint function conditions

The constraint function $C\colon \langle \mathsf{Type}, \sqsubseteq \rangle \to \mathcal{F}$ obeys the following conditions:

Type For a given type t, if $C(t)$ is the TFS $\langle Q, q_0, \delta, \theta \rangle$ then $\theta(q_0) = t$.

Monotonicity Given types t_1 and t_2, if $t_1 \sqsubseteq t_2$ then $C(t_1) \sqsubseteq C(t_2)$

Compatibility of constraints For all $q \in Q$, the TFS which is the substructure rooted at q, $F' = \langle Q', q, \delta', \theta' \rangle \sqsubseteq C(\theta(q))$ and $Feat(q) = Appfeat(\theta(q))$.

Maximal introduction of features For every feature $f \in \mathsf{Feat}$ there is a unique type t such that $f \in Appfeat(t)$ and there is no type s such that $t \sqsubset s$ and $f \in Appfeat(s)$.

3.5.10 Exercises

1. What would be the full constraint on **agr-phrase** if it was defined as shown in 3.53? What would the revised description for s_rule look like?

2. Considering the corrected animal hierarchy, Figure 3.14, augmented as described in §3.5.6. Write type definitions which would result in the full constraints described. Could this situation arise without there being any local constraint on **whale**?

3. Suppose I change the definition of **phrase** in grammar g1cfg to:

```
phrase := syn-struc &
[ ORTH *top*,
  ARGS *list* ].
```

What else would I have to do to the type system to keep it valid?

3.5.11 Answers

1. The full constraint would be the following:

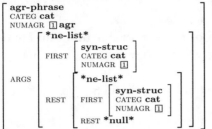

The revised description for s_rule would be:

```
s_rule := phrase &
[ CATEG s,
  ARGS [ FIRST [ CATEG np ],
         REST [ FIRST [ CATEG vp ]]]].
```

2. One possible set of constraint definitions is:

```
swimmer := animal &
[ FINS boolean ].
```

```
vertebrate := animal &
[ FRIENDLY boolean ].
```

```
vertebrate-swimmer := vertebrate & swimmer &
[ HARPOONED boolean,
  FINS true ].
```

so there need be no local constraint on **whale**.

3. Adding the feature ORTH to the type constraint for **phrase** would mean that ORTH was introduced at two places in the type hierarchy: **phrase** and **word**, thus violating the maximal introduction of features condition. The simplest way to fix this is simply to add the ORTH to the type **syn-struc**, since **phrase** and **word** both inherit from **syn-struc**. This would actually make the specification of ORTH on the type description for **phrase** redundant, but the type system would be formally valid.

3.6 Summary

This chapter has given a complete definition of type hierarchies, typed feature structures, subsumption, unification, type constraints and well-formedness conditions. The exercises in this chapter were designed to help you become thoroughly familiar with the typed feature structure formalism. In the course of the chapter, you should also have got some idea of how grammars work, and what the difference is between a simple context free grammar and a grammar that uses TFSs. The next chapter goes on to look at the use of TFSs in grammars in much more detail.

4

Grammars in typed feature structures

The previous chapter gave a complete description of the typed feature structure formalism. TFSs can be used to model things other than linguistic entities, but since our interest is in grammar modelling, I now turn to describing in more detail how grammars actually work. I'll start off with a discussion of how grammar rules and lexical entries are utilized during parsing and generation, first giving a rather abstract formal description and then briefly describing the parsing algorithm used in the LKB. (The discussion of generation is purely schematic, since a realistic discussion of generation requires that we know something about how semantics can be encoded in TFS grammars, which is left to §5.4.) Then I'll discuss the way that the LKB system processes files containing definitions of types, grammar rules, lexical entries and so on, and go into a lot more detail about the description language than in the previous chapter.

4.1   An introduction to grammars in TFSs

To define how parsing or generation work, in addition to the grammar definition in terms of TFSs, we need two extra pieces of information which are specified by *LKB parameters*.

1. The location of orthographic information in lexical structures (ORTH in `glcfg`). This is needed in parsing so that, when given an input string, the system can locate a set of lexical structures which correspond to the words in the string. Conversely, in generation, this enables the system to 'read off' a string from a list of lexical structures. The LKB parameter that specifies the path for orthographic information in lexical structures is `*orth-path*`. (In some of the grammars we'll look at in this book, the value of `*orth-path*` is

not simply ORTH. The value is set in the file globals.lsp for each grammar, as discussed later in this chapter.)

2. The location of the daughters and the mother structure in grammar rules. In the grammars we have looked at so far, the daughters are the elements of the ARGS list and the mother is the entire structure. The definition of a valid phrase, given below, depends on knowing the location of the mother and daughters. This value is also set by an LKB parameter, but this is kept constant for all the grammars we'll look at.

Besides specifying the location of the structures representing daughters in the grammar rule, the LKB parameters also specify the linear order in which daughters appear. This order is assumed to be fixed for each grammar rule. The fixed order assumption was made without much discussion at the beginning of the previous chapter, where it was simply carried over from the simple CFG. Fixed order is not, in fact, a necessary assumption for grammars using TFSs. However, since it makes the definition of a valid phrase simpler, and since many parsing algorithms, including those in the LKB, depend on this condition, I will assume it here. Similarly, I will assume that the number of daughters associated with each grammar rule is fixed, although again this is not actually a necessary assumption for TFS formalisms. The terminology used is *unary rule* for rules with one daughter, *binary rule* for two daughters, *ternary rule* for three and so on. In practice, grammars don't generally go much above ternary rules, and some grammars only use unary and binary rules. (The grammars we've looked at in detail so far only have binary rules, of course.)

At this point, it is useful to introduce the term *sign* to refer to a pairing of a structure and a string. For now, I will define a sign as a pairing of a TFS and a list of strings, with each string corresponding to a single word (in §4.3, below, we will see that we can alternatively encode strings as part of the TFS). For instance, the sign corresponding to *the dog* in grammar g2agr is as follows:

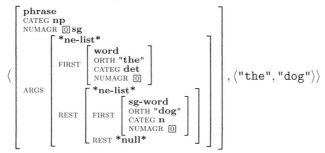

In the case of lexical entries in **g2agr**, the value of ORTH in the lexical entry TFS is equal to the single member of the list of strings that it is paired with.

The input to a parser or the output from a generator can be regarded as a list of word strings. For instance *the dog sleeps* is treated as ⟨"the", "dog", "sleeps"⟩. The process of converting the actual input to a list of word strings is *tokenization* and it can be quite complex. The tokenization function can be varied: however, in this book, we'll simply assume that word strings are delimited by spaces. A lexical lookup operation associates strings with signs. Lexical lookup for parsing takes a list of word strings as input and returns a list of sets of lexical signs, where each sign is a TFS paired with a list of a single word string. The list of sets of signs found for "the" "dog" "sleeps" in the grammar with agreement (**g2agr**) is shown below:

$$
\left\langle \left\{ \left\langle \begin{bmatrix} \textbf{word} \\ \text{ORTH "the"} \\ \text{CATEG } \textbf{det} \\ \text{NUMAGR } \textbf{agr} \end{bmatrix}, \langle\text{"the"}\rangle \right\rangle \right\}, \left\{ \left\langle \begin{bmatrix} \textbf{sg-word} \\ \text{ORTH "dog"} \\ \text{CATEG } \textbf{n} \\ \text{NUMAGR } \textbf{sg} \end{bmatrix}, \langle\text{"dog"}\rangle \right\rangle \right\}, \right.
$$

$$
\left. \left\{ \left\langle \begin{bmatrix} \textbf{sg-word} \\ \text{ORTH "sleeps"} \\ \text{CATEG } \textbf{vp} \\ \text{NUMAGR } \textbf{sg} \end{bmatrix}, \langle\text{"sleeps"}\rangle \right\rangle \right\} \right\rangle
$$

We have to think about sets of signs because of lexical ambiguity: a valid parse must involve exactly one structure from each set (though, for the grammars we have seen so far, there is no lexical ambiguity). For now, we will simply assume that lexical lookup adds a sign corresponding to a lexical entry to the set for a word string if the value of its ORTH feature matches the string. However, in realistic grammars, lexical lookup is complicated by morphology (discussed in the next chapter) and *multiword entries* (i.e., lexical entries corresponding to multiple units returned by the tokenizer, briefly discussed in §8.5).

We can also talk about an initial set of lexical signs for generation, but this is less intuitive. We have to regard every lexical sign in the grammar as a potential member of the initial set, since we do not have an initial sentence string.[23]

Besides the lexical signs, there is also a start structure, which is a TFS corresponding to the start symbol in a standard CFG.[24] In many grammars, a start structure for parsing will be specified by the grammar writer, as in the grammars we have seen so far. (If no start structure is

---

[23]Some generation algorithms, including that implemented in the LKB, do actually work from lexical signs, using the input semantics to filter the initial set for the generator, but I'll leave discussion of that until §5.4.

[24]The term *root condition* is sometimes used to mean roughly the same thing as start structure.

specified in a grammar, it is simply taken to be the most general TFS:
$\left[\,\text{*top*}\,\right]$.) In parsing, the string associated with the start structure
corresponds to the string for the complete sentence. So if we are parsing
⟨"the", "dog", "sleeps"⟩ with g2agr the start sign is:

$$\left\langle \begin{bmatrix} \textbf{phrase} \\ \text{CATEG s} \\ \text{NUMAGR } \textbf{agr} \\ \text{ARGS } \textbf{*list*} \end{bmatrix}, \langle \text{"the"}, \text{"dog"}, \text{"sleeps"} \rangle \right\rangle$$

For generation, the string in the start sign is initially unknown. However, when generating with larger-scale grammars, information about
semantics augments the start structure TFS.

Each phrase in parsing or generation is licensed by a grammar rule
and is associated with a list of strings which is formed by concatenating
the lists of strings associated with its daughters in a fixed order, which in
the grammars we have seen corresponds to the order of the elements on
the ARGS list of the rule. Parsing or generation then essentially involve
building as many phrases as are needed to link the lexical signs to the
start sign. To make this a little more precise, let's say that a *phrasal
sign* is a pair of a well-formed TFS F and a list of strings, such that F
is subsumed by some grammar rule and has structures in its daughter
positions which are each either subsumed by a member of the set of
lexical signs or are themselves valid phrasal signs. We then count as a
solution any phrasal sign which is also subsumed by the start structure
and which has the same string list as it does, if that is specified (i.e., the
string list for the sentence if we are parsing).

We can sum all this up as follows:

Properties of a grammar

> **Grammar** A grammar consists of a set of grammar rules,
> G, a set of lexical entries, L, and a start structure, Q,
> which are all well-formed TFSs. Each grammar rule in
> G has one or more daughter paths, $D_1 \ldots D_n$.
>
> **Lexical sign** A lexical sign is a pair $\langle L, S \rangle$ of a well-formed
> TFS L and a string list S, such that L is a lexical entry
> that matches the string S (S is L's ORTH value).
>
> **Valid phrase** A valid phrase P is a pair $\langle F, S \rangle$ of a TFS F
> and a string list S such that either:
>
> > 1. P is a lexical sign, or
> > 2. F is subsumed by some rule R and there are valid
> >   phrases $\langle F_1, S_1 \rangle \ldots \langle F_n, S_n \rangle$ such that the values
> >   of each of R's daughter paths $D_1 \ldots D_n$ subsume
> > $F_1 \ldots F_n$ respectively and S is the ordered concate-
> >   nation of $S_1 \ldots S_n$.

Sentences A string list S is a well-formed sentence according to the grammar if there is a valid phrase $\langle F, S \rangle$ such that the start structure Q subsumes F.

For the example sentence, *the dog sleeps*, there is only one solution, shown in 4.59. This corresponds to the application of the s_rule and the np_rule.

(4.59)

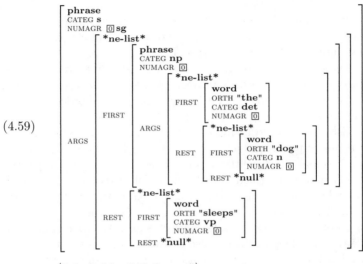

$\langle$"the", "dog", "sleeps"$\rangle$

Figure 10 shows the subsumption relationships involved in this analysis. The structures are arranged so that the lines between structures indicate a subsumption relationship, with the higher structure subsuming the lower one. The structure at the bottom corresponds to the structure for the complete parse, as shown in Figure 4.59, although to save space, the type **ne-list*** has been omitted here and I have omitted the subsumption relationship with the start structure. The ovals indicate the parts of the structures which are subsumed. By going through the figure, you should be able to see that the lowermost structure meets the criteria for well-formed TFSs in a phrasal sign.

1. it is subsumed by the structure for the S rule
2. its second daughter is subsumed by a lexical sign (that for *sleeps*)
3. its first daughter is a well formed phrasal sign because:
   (a) it is subsumed by the structure for the NP rule (the structure at the top left of the figure)
   (b) its first daughter is subsumed by the lexical sign for *the*
   (c) its second daughter is subsumed by the lexical sign for *dog*

Since this structure corresponds to the string for the complete utterance *the dog sleeps*, and is subsumed by the start structure (not shown here), it is a valid solution.

In principle, this notion of a validity being defined in terms of connecting a start structure with lexical structures via grammar rules holds equally well for parsing and generation. All that differs is the information we start with. However, for grammars like the ones we've been dealing with so far, the start structure and the lexical structures are too underspecified for practical generation. For these tiny grammars, because there is no recursion, the number of valid strings is finite, so we could generate all strings licensed by the grammar. But in general this is not possible because the number of strings is infinite. There is in fact nothing to stop us defining operations which are a mixture of parsing and generation. But generally, parsing is assumed to involve starting with an input string and constructing all valid structures which can be associated with that string, while generation involves starting with a structure and deriving all valid strings.

Properties of parsing and generation

Parsing a string, S, consists of finding all valid phrases $\langle F_1, S \rangle \ldots \langle F_n, S \rangle$ such that the start structure Q subsumes each structure F_1 to F_n.

Generating from a start structure, Q', which is equal to or subsumed by, the general start structure Q, consists of finding all valid strings, $S_1 \ldots S_n$ which correspond to valid signs, $\langle F_1, S_1 \rangle \ldots \langle F_n, S_n \rangle$ such that the start structure Q' subsumes each structure F_1 to F_n.

4.1.1 Derivations and underspecification

We can talk about a particular *derivation* of a structure for a sentence in terms of application of rules to signs. For instance, the derivation for the sentence shown in Figure 4.59 could be represented in terms of the identifiers for the lexical entries and rules that are involved, as follows:

```
(s_rule (np_rule the dog) sleeps)
```

However, the definition of valid phrases in terms of subsumption has the consequence that there may be multiple valid TFSs corresponding to a single derivation. For instance, suppose that we add the following lexical signs to grammar g2agr:

```
sheep := word &
[ ORTH "sheep",
  CATEG n ].
```

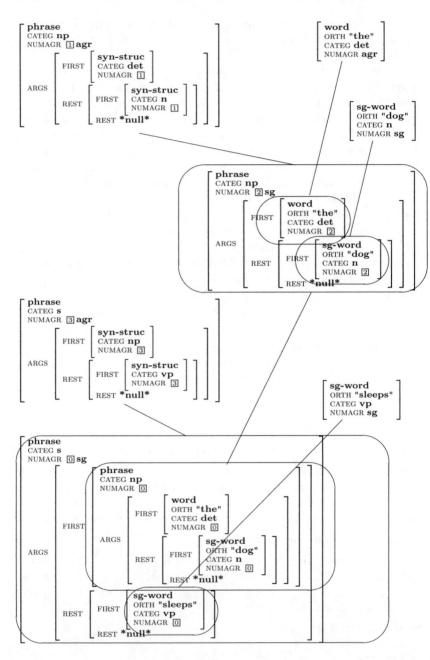

FIGURE 10 Parsing as subsumption

```
slept := word &
[ ORTH "slept",
  CATEG vp ].
```

By specifying the structures as having type **word** as opposed to **sg-word** or **pl-word**, I have *underspecified* the number for the lexical entry for *sheep* as I earlier did for the entry for *the*.[25] We then end up with multiple valid structures for a single derivation of the sentence *the sheep slept*, as shown in Figure 11 (I have omitted the type node label ***ne-list*** to save space). The difference between these structures is the value of NUMARG, which is **agr** in the first structure, **sg** in the second and **pl** in the third.

The second and third structures in Figure 11 are valid according to our definition, but they contain extra information that could not have come from the rules or lexical entries. At least for practical purposes, we want a parse to result in a single structure for a particular derivation, and we don't want that structure to contain any extra information, especially since, with more complex grammars, we could end up with an infinite number of valid phrases corresponding to the sentence. For a particular derivation, there will always be a single structure which subsumes all the other valid structures (e.g., the top structure in Figure 11), so this is the structure that we want the parser to return for that derivation.

Contrast this situation with the one where *sheep* is lexically ambiguous. That is, instead of the entry as shown above, we have two entries corresponding to the single string "sheep":

```
sheep_sg := sg-word &
[ ORTH "sheep",
  CATEG n ].
```

```
sheep_pl := pl-word &
[ ORTH "sheep",
  CATEG n ].
```

```
slept := word &
[ ORTH "slept",
  CATEG vp ].
```

In this case, we will end up with two derivations for the sentence:

---

[25]This is a somewhat controversial thing to do for *sheep*. There are arguments that it would be better to treat it as being lexically ambiguous: i.e., that the string "sheep" should correspond to two lexical entries, one **sg-word** and one **pl-word**. I'll assume the underspecification account here for purposes of illustration, but I'll contrast it with the ambiguity account below.

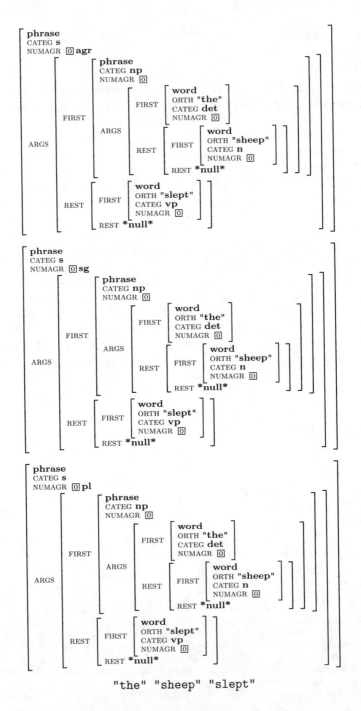

"the" "sheep" "slept"

FIGURE 11 Structures for a sentence underspecified for agreement

```
(s_rule (np_rule the sheep_sg) slept)
```

```
(s_rule (np_rule the sheep_pl) slept)
```

For the first of these derivations, the second structure in Figure 11 is valid, for the second derivation, the third structure is valid. The top structure is not valid in either derivation, because its value for NUMARG means that part cannot be subsumed by any lexical entry.

Returning to the underspecified entry for *sheep*, it may seem counterintuitive to regard the structures which contains extra information about number as valid. I could, in fact, have defined the notion of validity so that there was only one valid structure for each possible derivation: that is, the most general one that fit the subsumption conditions. This is guaranteed not to contain any extra information. However, this would mean that different notions of valid structures corresponding to a particular string were necessary for parsing as opposed to generation. For instance, a generator might have an input specification such that NUMARG had **sg**, but we do not want to claim that this structure is invalid, even if the only string in the grammar that could realize it is also compatible with NUMARG being **pl**. A parser that only explicitly returns one structure per derivation (such as the LKB parser) can still be regarded as meeting the definition of parsing that I gave above, since the other valid structures could be constructed automatically from the returned structure (without knowledge of the grammar).

4.1.2 Parse trees

As I mentioned briefly in the previous chapter, in a TFS grammar, the idea of a *parse tree* is just an abbreviation for the complete representation of a structure corresponding to the valid parse. The TFS shown in Figure 4.59 might be abbreviated by just taking the symbols for the values of CATEG, as shown below:

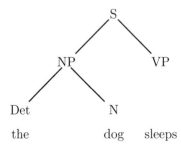

This is indeed what you will see as the parse tree for the sentence in the LKB with grammar **g2agr**. In the LKB system, the trees are simply convenient abbreviations so that a grammar writer can quickly get an idea of the analyses of a sentence. The grammar writer provides definitions that specify which node labels correspond to which TFSs, as is explained in §4.5.7.

It is important to understand that a tree represents the complete phrase for a sentence, not the procedure by which a parse is constructed. If you click on the node corresponding to the entry **the** in the enlarged tree for the parse of *the dog sleeps* in the LKB, you will see that the TFS associated with it has an instantiated value for NUMARG. Similarly, if we defined tree labels to be a concatenation of values for CATEG and NUMARG, which is effectively what has been done in grammar **g3nodes**, the tree would have a label Det-sg, rather than simply Det, as shown below:[26]

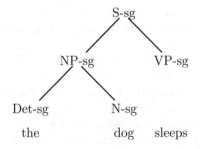

This effect is because the node labelled Det-sg in the tree corresponds to part of the final TFS, so it is subsumed by the entry **the**, not identical to it. (However, the LKB also provides a method for looking at the intermediate structures in the construction of a parse, the parse chart, as discussed in §4.2, below.)

4.2   Parsing in an implementation

Although parsing in the LKB formally corresponds to the description given above, for practical purposes we need something rather more efficient.[27] The first necessity is to set up parsing in terms of unification rather than subsumption. As discussed above, the parser returns the most general structure that meets the subsumption conditions, and this corresponds to the unification of the relevant structures. For instance, a structure that corresponds to the phrase **"the dog"**, is the unification

---

[26]The labels are not literally constructed by concatenation: the exact mechanism of the parse tree node labelling will be discussed in §4.5.7.

[27]This chapter and the next contain several remarks about efficiency, all of which are to be understood informally because they are based on experience with practical systems and not theoretical complexity.

of the relevant parts of the NP rule with the lexical TFSs. This is shown below (recall also the examples in §3.4.1):

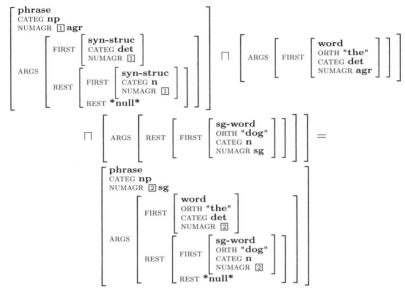

Note that I have shown the TFSs corresponding to the daughters as embedding the daughter TFSs themselves at the end of the appropriate path (i.e., ARGS.FIRST for the first daughter and ARGS.REST.FIRST for the second daughter.).

Another unification could combine this result with the sentence rule plus the lexical sign for *sleeps* as shown below:

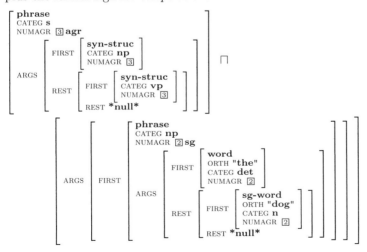

$$\sqcap \left[\text{ARGS} \left[\text{REST} \left[\text{FIRST} \begin{bmatrix} \textbf{sg-word} \\ \text{ORTH "sleeps"} \\ \text{CATEG } \textbf{vp} \\ \text{NUMAGR } \textbf{sg} \end{bmatrix} \right] \right] \right]$$

The primary reason why it is useful to think in terms of unification rather than subsumption is that efficient algorithms exist for unifying two structures. I defined the unification of two structures as the most general TFS which they mutually subsume, but unification can be implemented directly, without computing all the subsumptions (which would be massively inefficient). Thus, rather than searching for all TFSs subsumed by the structure for **np_rule** with daughters subsumed by the relevant lexical entries, the parser performs a deterministic unification procedure directly in order to build the relevant structure. If a unification step fails, then this is because of some incompatibility among the structures such that the grammar rule is not applicable. This may be because of incompatibility between one of the daughters and the rule, or, even if each daughter is pairwise compatible with the rule, the combination of all of them together may be inconsistent.[28]

The current LKB unification algorithm is described in Malouf et al (2000) and some other references to books and papers that discuss unification algorithms are given in §5.8, but luckily grammar engineers don't actually need to know about the implementation of unification. However, it is useful to know that the computational cost of unification is proportional to the size of the TFS and in grammars which manipulate large TFSs, unification can be quite expensive. Most implementations of parsing that use unification, including that in the LKB, require TFSs to be copied sometimes, which is one reason that large LKB grammars need machines with substantial amounts of memory. Some of the efficiency techniques described in Chapter 8 concern ways of testing rules for applicability without doing full unifications, while others reduce the size of the TFSs under consideration.

Unification provides a reasonably efficient way of testing rules for applicability, and of actually applying them. The other requirement for an implementation is to control the search space so that the parser does not have to try all rules and all phrases. We need a strategy for searching the space of valid phrases and a way of recording the parts of the space that have been explored. Given the assumptions that we have made about fixed arity of rules and fixed daughter orders within a rule, it is possible to adapt well-known algorithms from non-unification

---

[28]Tools are available for debugging grammars in the LKB system that allow the unifications to be carried out one step at a time, under the control of the user, as discussed in detail in §7.4.1.

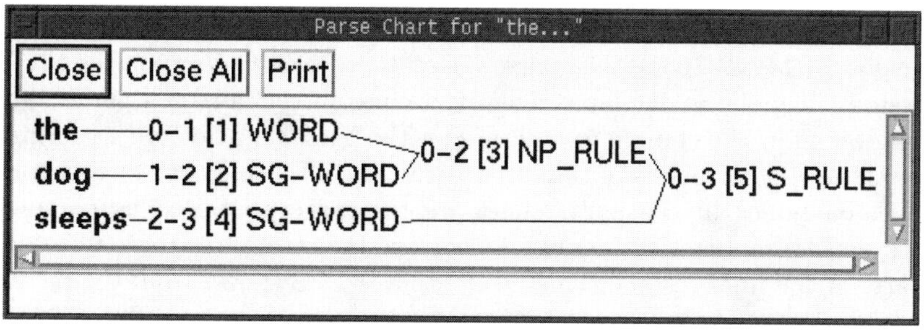

FIGURE 12 LKB parse chart for *the dog sleeps*

based formalisms, such as simple CFGs. The approach used in the LKB is *chart parsing*, a technique which has been used in a great many systems. I will not go through chart parsing in detail here since it is so well known: see Jurafsky and Martin (2000) for a full discussion, for instance. The grammar engineer does not need to know the full details of the parsing algorithm, but a basic understanding is useful. The *parse chart*, or simply *chart*, is the data structure which records the partial results, known as *edges*. Edges correspond to valid signs, with additional information about the parts of the input string that they span. For instance, in *the dog sleeps*, the edge for *the dog* contains the sign, plus the start vertex 0 and the end vertex 2, while the edge for *sleeps* would have a start vertex 2 and an end vertex 3 (this can be seen in the LKB parse chart display, described below).

In the simplest version of chart parsing, processing proceeds *bottom-up*. After the initial lookup to find the lexical signs associated with a string, the parser combines lexical structures together to form phrases, and then combines these phrases with one another, or with other lexical structures. The information in the start structure is not checked till the end. At each point, the structures that are built are stored on the chart, so that the same phrase never has to be constructed twice. The LKB's current parsing algorithm is essentially bottom-up: it is described in Oepen and Carroll (2000b). Although it is an active chart parser, the edges shown on the display are all passive edges.

To open a window displaying a parse chart in the LKB, you can either click on the **Show chart** button on the small parse tree display, or on **Show parse chart** from the **Parse** menu in the LKB top window. The parse chart window shows all the edges that are constructed during an attempt to parse a string. The chart for *the dog sleeps* is shown in Figure 12. A more detailed description of the LKB parse chart display is in §6.6.

There is a difference between the parse chart and the parse tree display of TFSs. To see this, parse the sentence *the dog sleeps* in the LKB system. The TFSs displayed when you click on the enlarged parse tree window show the complete structure for the parse (the bottom structure in Figure 10), or pieces of that structure. The TFSs displayed when you click on nodes in the parse chart are the structures used as parsing progresses (in terms of Figure 10, they are the structures that subsume parts of the final result). This means that a TFS shown from the parse tree display may be more specific than the corresponding TFS in the parse chart display. For instance, the structure at the end of the path ARGS.FIRST.ARGS.FIRST in the TFS at the bottom of Figure 10 can be displayed from the parse tree display in the LKB by clicking on the Det node and selecting **Feature Structure** from the pop-up menu. It is more specific with respect to the value of NUMAGR than the lexical entry for *the*, which was shown in the top left corner of Figure 10. This latter structure is the one which is shown from the parse chart window, as you can see by clicking on the **word** edge that spans 0-1 and selecting **Feature Structure** from the menu.

The parse tree display is primarily used for getting an overview of the results of parsing while the parse chart is more useful for detailed debugging, because the TFSs are simpler and because the unification checking tool (described in §7.4.1) works more naturally in conjunction with the parse chart.

It is extremely important to realize that the LKB's processing strategy does not determine the results of parsing, provided that parsing proceeds to completion. Alternative parsing strategies would give exactly the same results. It is useful for the grammar engineer to know a little about parsing strategies, because the parse chart is a helpful debugging tool, and because knowing something about how the parser works is necessary for writing really efficient grammars. But grammars written in the LKB could always be processed with a different strategy.

4.2.1 Exercise

The grammars we have looked at in detail so far are too simple to really demonstrate the utility of chart parsing. In this exercise, we add a very simple rule to the grammar which makes it recursive and which introduces ambiguity.

1. Add descriptions for the nouns *lion, hunter, opponent* and *mountain* to the lexicon in g2agr.
2. Add a rule for noun-noun compounding, analogous to the CFG rule N -> N N.

3. Save your files, reload the grammar and then parse the following sentences, noting the number of parses and the number of edges in the chart[29]:

   (a) the mountain lion sleeps

   (b) the mountain lion hunter sleeps

   (c) the mountain lion hunter opponent sleeps

4. Add lexical descriptions for *roar*, *roars*, *echo* and *echoes* as both verbs and nouns. The different entries for the same word must have distinct identifiers: for instance **roar_v** and **roar_n**. Save, reload and parse:

   (a) the mountain lion roars echo

Examine the parse trees and the parse chart, noting edges which do not contribute to the final parse.

4.2.2 Answers

A grammar with the required changes is shown in **an5**.

4.3 Difference lists

I now return to the issue of how we can treat strings as part of a TFS rather than having to define signs so that strings are external. Of course, in the grammars we have seen so far, the string value of a lexical entry is stored in ORTH. Why can't we do this for phrases too? It looks like we could make a binary phrase have an ORTH which contains two features, ORTH1 and ORTH2 which are coindexed with the daughters' ORTH values. For example, the following would be the structure for the ORTH of *the dog sleeps*:

$$
\begin{bmatrix} \text{ORTH} & \begin{bmatrix} \text{ORTH1} & \begin{bmatrix} \text{ORTH1} & \texttt{"the"} \\ \text{ORTH2} & \texttt{"dog"} \end{bmatrix} \\ \text{ORTH2} & \texttt{"sleeps"} \end{bmatrix} \end{bmatrix}
$$

But this is not what we want, because it contains more structure than we actually know from the initial string. Effectively the ORTH is encoding the rule application structure: the structure could be different for a different analysis or a different grammar, so this won't work as the input to a parser. Ideally we want to simply append the ORTH structures

---

[29]Note for those who are familiar with chart parsing: when parsing with a conventional CFG, only one edge is added to the chart for each constituent of any category that spans a given part of the input string. With unification based grammars, we can have multiple structures with the same category, but with different derivations. These are treated as distinct in the LKB parser, with the effect that there can be an exponential number of edges, even for a grammar that is equivalent to a conventional CFG, as this one is. There are versions of the LKB parser that get round this, as discussed briefly in §8.3.4.

together (i.e., concatenate them so we have a simple list as a result). But we cannot define an append of two arbitrary lists in the LKB's TFS formalism, because append requires that we know where the end of the first list is. We can easily build a phrase whose orthography is the append of two lists of known length. For instance, the structure below appends a list of length two with another list:

$$
\begin{bmatrix}
\text{ORTH} & \begin{bmatrix} \text{FIRST } \boxed{1} \\ \text{REST} & \begin{bmatrix} \text{FIRST } \boxed{2} \\ \text{REST } \boxed{3} \end{bmatrix} \end{bmatrix} \\
\\
\text{ARGS} & \begin{bmatrix} \text{FIRST} & \begin{bmatrix} \text{ORTH} & \begin{bmatrix} \text{FIRST } \boxed{1} \\ \text{REST } \begin{bmatrix} \text{FIRST } \boxed{2} \end{bmatrix} \end{bmatrix} \end{bmatrix} \\ \text{REST} & \begin{bmatrix} \text{FIRST} & \begin{bmatrix} \text{ORTH } \boxed{3} \end{bmatrix} \end{bmatrix} \end{bmatrix}
\end{bmatrix}
$$

But this only works because of the reentrancies/coindexations and these would have to point to different paths for lists of different lengths. Since we want to append lists of arbitrary lengths, this won't work either.

The standard solution to this problem is to maintain pointers to the last element in a list. Structures with such pointers are known as *difference lists*. We can implement difference lists by means of a type which has two appropriate features, LIST and LAST, both of which take a value which is of type *list*.

```
*diff-list* := *top* &
[ LIST *list*,
  LAST *list* ].
```

When we use this type for actual difference lists, the value of LIST is a non-terminated list, that is, one whose final REST value is a *list* rather than a *null*. The value of LAST is coindexed with the final REST value. For instance, the following is a valid difference list (assuming **a**, **b** and **c** are valid types):

$$
\begin{bmatrix}
\textbf{*diff-list*} \\
\text{LIST} & \begin{bmatrix} \text{FIRST } \textbf{a} \\ \text{REST} & \begin{bmatrix} \text{FIRST } \textbf{b} \\ \text{REST} & \begin{bmatrix} \text{FIRST } \textbf{c} \\ \text{REST } \boxed{1} \end{bmatrix} \end{bmatrix} \end{bmatrix} \\
\text{LAST } \boxed{1}
\end{bmatrix}
$$

A grammar that includes difference lists for orthography is in the directory g4diff. I have added the type *diff-list* to the type system in g2agr, replaced **syn-struc** with **sign**, and redefined **word** and **phrase** as follows:

```
sign := *top* &
[ ORTH *diff-list*,
  CATEG cat,
  NUMAGR agr ].

phrase := sign &
[ ORTH [ LIST #first,
         LAST #last ],
  ARGS [ FIRST [ ORTH [ LIST #first,
                        LAST #middle ] ],
         REST [ FIRST [ ORTH [ LIST #middle,
                               LAST #last ]]]]].
word := sign &
[ ORTH [ LIST [ FIRST string,
                REST #end ],
         LAST #end ]].
```

I also had to redefine the individual lexical descriptions in g4diff so their orthography goes in the right place in the difference list, for instance:

```
this := word &
[ ORTH [ LIST [ FIRST "this"]],
  CATEGORY det ].
```

Experiment with parsing some sentences with g4diff and look at the values for the ORTH of the phrases. Notice also the values for ORTH of each of the lexical signs as shown from the parse tree display. If this seems odd, remember that the parse tree shows the structures which make up the complete structure for the sentence, as opposed to the parse chart, which displays the original structures of the lexical entries. The value for ORTH.LIST for each lexical entry is a non-terminated list: the orthography values for any words that come after the lexical sign you are looking at subsume its ORTH.LIST.REST in the structure for the sentence as a whole.

The use of difference lists to encode orthography allows us to define signs as TFSs, since the TFS itself can include a representation of the string for any phrase. This is not actually necessary (or even particularly useful) for the operation of the current LKB parser, but difference lists are required to build up other types of list, some of which are exemplified in the grammars we will discuss in the next chapter. Difference lists are used for encoding semantics and also for unbounded dependencies.

4.3.1 *Alternative definitions of a grammar

This section is a digression, concerning alternative formal definitions of well-formedness of phrases that are available once we encode orthography as part of the phrasal TFSs. It does not have any relevance for parsing with the current LKB system, so can be skipped if preferred.

The way we used difference lists in the previous section, to allow the orthography to be built up on phrases, means that the grammar itself encodes the correspondence between phrases and strings. So we can define a grammar without having to talk about strings as being external to the TFSs. A TFS for a phrase is itself a sign, because all valid phrases are TFSs which have an orthography value which is a string. We also don't have to talk about concatenation of strings explicitly, because the type constraint on the type **phrase** enforces the effect that the concatenation of strings on the daughters is equal to the orthography of the phrase as a whole. If TFSs represent complete signs and concatenation of strings is encoded in the grammar, we have the following revised descriptions of a valid phrase and of a well-formed sentence:

Revised properties of a grammar, encoding orthography in TFSs

> **Valid phrase** A valid phrase P is a well-formed TFS such that either:
>> 1. P is a lexical sign, or
>> 2. P is subsumed by some rule R and there are valid phrases $F_1 \ldots F_n$ such that the values of each of R's daughter paths $D_1 \ldots D_n$ subsume $F_1 \ldots F_n$ respectively.
>
> **Sentences** A string S is a well-formed sentence according to the grammar if there is a valid phrase F with an orthography value of S such that the start structure Q subsumes F.

This simplifies the definition somewhat, but there is a still neater definition which connects the notion of well-formedness of phrases to that of well-formedness of particular types. We assume that all signs are TFSs which are of type **sign** or one of its subtypes. The types of the nodes at the end of any daughter paths in a sign are similarly constrained to be subtypes of **sign**. Then we have the following:

Grammars as additional constraints on the type sign

> All well-formed TFSs of type **sign** must be subsumed by a grammar rule R or a lexical entry L.
>
> A string S is a well-formed sentence according to the gram-

mar if there is a well-formed TFS of type **sign** which is subsumed by the start structure Q and has an orthography value of S.

This gives the same effect as the previous definition: the notion of well-formedness recursively affects the nodes in a phrase that represent its daughters since these are also of type **sign**.

This definition is quite a bit more elegant than the one in §4.1 because we have moved most of the work into the type system. We have gone from a definition of grammar that required various assumptions about orthography position, daughter paths and string concatenation, to one that simply requires that we know the set of lexical entries and the set of grammar rules. What's more, under this approach, parsing and generation are general enough to include grammars which do not make the assumptions about word order and concatenation that we started off with. All that is required is to change the grammar so that phrases build up their orthography in a different way.

However, unfortunately this definition has moved some considerable way away from the version of parsing used in the LKB. Although future versions of the LKB may parse grammars which do not assume strict string concatenation, the current parsing algorithm requires that the string order mirrors the order in the ARGS list.

4.4 The description language

Having discussed how grammars work, I now go into more detail about how to write a grammar in the LKB. In this section, I discuss the description language – in the following sections I will discuss the organization of files that contain the descriptions and how they are processed by the LKB.

In order to specify types, constraints, lexical entries and so on, we require a *description language*, which I introduced informally in the previous chapter. Multiple description languages are supported by the LKB, but the one we use in this book is syntactically based on the Type Description Language (TDL) originally used in the DISCO/PAGE system (Krieger and Schafer, 1994). The syntax used in the LKB is actually the subset of TDL used in the LinGO ERG, with one or two LKB-specific extensions, and an LKB description gives different behaviour from a PAGE description, because of differences in the underlying formalism: I will nevertheless just refer to the description language as TDL.

The purpose of a description language is to allow the grammar designer to write definitions of types and of TFSs in a way which is readable by a machine. A valid description always results in a single TFS. Exactly the same grammar could be described using different description languages. As discussed so far, TDL looks like an ASCIIised version of the AVM syntax. However, description languages generally allow various abbreviations. One useful abbreviation allows lists to be encoded without explicitly mentioning FIRST and REST. In TDL, lists begin with a <, end with a > and , separates list elements. For instance, the following two descriptions are equivalent:

```
example := phrase &
[ CATEG np,
  ARGS  < word &
          [ ORTH "these",
            CATEG det ],
          syn-struc &
          [ CATEG n ] > ].


example := phrase &
[ CATEG np,
  ARGS  *ne-list* &
          [ FIRST word &
                  [ ORTH "these",
                    CATEG det ],
            REST  *ne-list* &
                  [ FIRST syn-struc &
                          [ CATEG n ],
                    REST *null* ]]] .
```

I will introduce the syntax of TDL in detail by going through some examples, in increasing order of complexity. After this, I give the formal definition of the syntax.

4.4.1 Syntax of type and constraint descriptions
Example 1

```
cat := *top*.
```

This defines the type **cat** to inherit from the single parent ***top***, i.e., ⊤. Since no constraint is specified for **cat** and it is not possible to specify a constraint for ***top***, the constraint on **cat** is simply [cat]. Note that:

1. The type definition is terminated by a full stop ('.') — in fact all TDL descriptions are terminated like this.

2. New lines and spaces are not significant, except that spaces may not occur in identifiers, such as type names, names of lexical entries, features etc. But it makes the files much more readable if a

consistent approach to indentation is maintained.

3. Case is not significant, though there is a convention that features are written in uppercase and types in lowercase.

4. The use of * in the type name ***top*** and so on is a convention indicating that these are standard types which are common to many grammars. However the *s are treated as normal characters and there is no formal difference between these types and the others.

5. The following characters may not be used in identifiers:
   < > ! = : . # & , [ ] ; @ $ ( ) ^ "

6. Comments are allowed in LKB files, but only in between descriptions. Comments are either single lines beginning with ; or they are bracketed by #| |#.

Example 2

```
*ne-list* := *list* &
 [ FIRST *top*,
   REST *list* ].
```

The constraint feature structure comes after the parent type and is enclosed by [ ]. Note also the comma between the two attribute-value pairs. This description defines ***ne-list*** to inherit from ***list*** and to have a constraint specification which is equivalent to the following in the 'pretty' AVM notation:

$$\begin{bmatrix} \text{FIRST} & \textbf{*top*} \\ \text{REST} & \textbf{*list*} \end{bmatrix}$$

The actual constraint on the type will be the following:

$$\begin{bmatrix} \text{FIRST} & \textbf{*top*} \\ \text{REST} & \textbf{*list*} \end{bmatrix} \sqcap C(\textbf{*list*})$$

Example 3

```
head-feature-principle := phrase & headed &
 [ HEAD #head,
   HEAD-DTR [ HEAD #head ]].
```

This descriptions states that the type **head-feature-principle** has two parents: **phrase** and **headed**.[30] The embedded TFS is enclosed by [ ], just like the top-level TFS. The structure shows the notation for reentrancy/coindexation: the reentrant node is indicated by the tag #head: tags always begin with #. Note that the name of the tag is not significant outside the immediate context of the description (e.g., instead of the two instances of #head above, we could have had two instances of #h or of #1, without affecting anything else in the grammar).

---

[30]This example does not correspond to a type in any of the sample grammars.

The constraint on this type will be:

$$\begin{bmatrix} \textbf{head-feature-principle} \\ \text{HEAD} \; \boxed{1} \; \textbf{*top*} \\ \text{HEAD-DTR} \; \begin{bmatrix} \text{HEAD} \; \boxed{1} \end{bmatrix} \end{bmatrix} \; \sqcap \; C(\textbf{phrase}) \; \sqcap \; C(\textbf{headed})$$

plus any extra constraints arising from the types of the substructures, since the TFS will be made well-formed, as discussed in §3.5.

Note that the TDL syntax also allows an alternative notation to be used for concatenation of features:

```
[ HEAD #head,
  HEAD-DTR.HEAD #head ].
```

is equivalent to:

```
[ HEAD #head,
  HEAD-DTR [ HEAD #head ] ].
```

Example 4

```
noun-lxm := lexeme &
[ HEAD noun & [ NUMAGR #agr ],
  SPR [ FIRST [ HEAD det & [ NUMAGR #agr ],
               SEM.INDEX #index ],
       REST *null*],
  COMPS *null*,
  SEM [ INDEX object & #index ] ].
```

This more complex example is included to show that when two pieces of information, such as a type and a coindexation tag, or a type and an embedded TFS, are specified for the same node, they are conjoined by &. For instance, the node reached by the path SPR.FIRST.HEAD has type **det** and embedded TFS [NUMAGR #agr]. Similarly SEM.INDEX has type **object** and coindexation tag #index.

It would actually be valid TDL to avoid the use of commas and to write:

```
*ne-list* := *list* &
 [ FIRST *top* ] &
 [ REST *list* ].
```

instead of:

```
*ne-list* := *list* &
 [ FIRST *top*,
   REST *list* ].
```

It is not normal to do so, because it is longer and a bit less clear. Similarly, the following is equivalent to the description of **noun-lxm** above, but does not use & or internal []s.

```
noun-lxm := lexeme &
[ HEAD noun,
  HEAD.NUMAGR #agr,
  SPR.FIRST.HEAD det,
  SPR.FIRST.HEAD.NUMAGR #agr,
  SPR.FIRST.SEM.INDEX #index,
  SPR.REST *null*,
  COMPS *null*,
  SEM.INDEX object,
  SEM.INDEX #index ].
```

Most people find this less easy to read, but if you find the TDL syntax very difficult to write, you can always resort to this style.

Example 5

```
sheep-relation := relation &
[ PRED 'sheep,
  ARG0 sement ].
```

Strings are are indicated by ' (e.g., 'this) or " (e.g., "this"). The " syntax can be used for strings with spaces, for instance:

```
ad-hoc-relation := relation &
[ PRED "ad hoc",
  ARG0 sement ].
```

Although these examples are syntactically valid, they are not part of a realistic grammar: on the whole, strings tend to only occur in the lexicon.

4.4.2 Description of lists and difference lists

Lists and difference lists are very frequently used constructs in many typed feature structure grammars. Because of this, the TDL syntax provides a notation for them directly, which expands out into the typed feature structure representation. In the LKB, the names of features and types involved are controlled by system parameters in order to allow this expansion (see §9.2.2). For example, suppose lists are defined as in the example grammars:

```
*list* := *top*.

*ne-list* := *list* &
  [ FIRST *top*,
    REST *list* ].

*null* := *list*.
```

Then the full representation of a list of elements a,b,c would be:

```
[ FIRST a,
  REST [ FIRST b,
         REST [ FIRST c,
                REST *null* ]]]
```

Using the abbreviatory notation, we could simplify this to:

```
< a, b, c >
```

Here and below, the a, b, c can be arbitrarily complex conjunctions of terms, which themselves contain lists etc.

It is often necessary to specify potentially non-terminated lists, for example:

```
[ FIRST a,
  REST [ FIRST b,
         REST [ FIRST c,
                REST *list* ]]]
```

which allows the possibility of additional elements after c. This is represented in the abbreviatory notation as

```
< a, b, c ...>
```

Finally, the syntax of TDL allows for *dotted-pairs* (although we won't use these in this book). For instance, the pair a.b could be represented as:

```
[ FIRST a,
  REST b ]
```

which can be abbreviated as:

```
< a . b >
```

Difference lists can also be abbreviated. For example, assume the following type definition:

```
*diff-list* := *top*
             [ LIST *list*,
               LAST *list* ].
```

The full notation for the difference list containing a,b,c would be:

```
[ LIST [ FIRST a,
         REST [ FIRST b,
                REST [ FIRST c,
                       REST #last ]]],
  LAST #last ]
```

This can be abbreviated as:

```
<! a, b, c !>
```

Note that the LIST component of the difference list is not terminated.
It is not possible to stipulate a difference list which has an arbitrary
number of elements, since the coindexation of LIST and LAST requires a
determinate path.

4.4.3 Descriptions of lexical entries

The syntactic description for lexical entries is very similar to that for
type descriptions. For example:

```
this_1 := sg-word &
[ ORTH [ LIST [ FIRST "this",
                REST #end ],
         LAST #end ],
  CATEG det ].
```

which could alternatively have been written:

```
this_1 := sg-word &
[ ORTH <! "this" !>,
  CATEG det ].
```

However the interpretation of lexical descriptions and type descriptions is different. Rather than stipulating that this_1 is a type in the
hierarchy under the type **sg-word**, this definition states that this_1 is a
TFS which has the type **sg-word** and identifier this_1. If the definition
were simply:

```
this_1 := sg-word.
```

this should be read as stating that the TFS named by this_1 is [**sg-word**]:
that is the single-node TFS with the node having type **sg-word** (though
obviously the process of making this well-formed expands the structure).
The distinction between types and entries is discussed further in §4.4.5.

Note that, although it is legal syntax to define a lexical entry with
multiple types on a node, this will not be valid unless the types have an
existing type as their greatest lower bound. Thus it is normal to specify
this type directly in lexical entries.

4.4.4 Descriptions of grammar rules

Grammar rules in the LKB system are TFSs, which represent relationships between two or more signs. Their descriptions are syntactically
identical to lexical entries.

4.4.5 The distinction between types and entries

One thing that it is important to be clear about is the distinction between the definitions of types and everything else (i.e., the entries). The type definitions package together two distinct pieces of information: the location of the type in the hierarchy, and the definition of the constraint on that type. So the following description of the type **a**:

```
a := b & c.
```

states that the type **a** is a lower bound of **b** and **c** in the hierarchy. Elsewhere in the TDL description language, & can be taken as equivalent to unification, but here it is understood as defining part of the type hierarchy. It follows from this type definition that the constraint on **a** will be the unification of the constraints on type **b** and type **c**, but as we've seen we need to know what the type hierarchy is before we can talk about unification, so the statement about the hierarchy is logically prior.

The same description used in an entry (e.g., in the description of a lexical entry or grammar rule) must be read quite differently. In the context of the description of an entry, it would mean that the TFS named **a** is the unification of the TFSs $\begin{bmatrix} b \end{bmatrix}$ and $\begin{bmatrix} c \end{bmatrix}$.[31] Since this is only valid if there is some type **d** which is the greatest lower bound of **b** and **c**, it isn't usual to write descriptions of entries like this, since it's clearer to simply write:

```
a := d.
```

4.4.6 *A formal description of the syntax of descriptions

To formally express the syntax of the description language, we use Backus-Naur Form (BNF), with mostly standard notational conventions: alternative right-hand sides of productions are indicated by |. However, I will follow the convention that both non-terminals and terminal tokens are italicized: non-terminals are capitalized. For example *Avm-def* is a non-terminal, *identifier* is a terminal token that could match a type name such as 'head-dtr-type'. Using these conventions, the following is the basic syntax of the type specification language:

---

[31]If you have looked at the formal notation used in Chapter 3, you will notice that I overloaded the same operator so it stands for both unification and greatest-lower bound in the type hierarchy. In that context, however, there is no ambiguity as to whether we are referring to types or feature structures. Unfortunately TDL does not distinguish between a type **t** and the TFS consisting of a single node labelled with that type, which we write as $\begin{bmatrix} t \end{bmatrix}$ in the AVM notation. This means that a syntactically identical description is understood quite differently if it is the description of a type as opposed to the description of an entry. As with QWERTY keyboards, we are now stuck with the convention.

Type-def → *Type Avm-def* .
Type → *identifier*
Avm-def → **:=** *Conjunction*
Conjunction → *Term* | *Term* **&** *Conjunction*
Term → *Type* | *string* | *Feature-term* | *Coreference* | *List* | *Diff-list*
Feature-term → **[]** | **[** *Attr-val-list* **]**
Attr-val-list → *Attr-val* | *Attr-val* **,** *Attr-val-list*
Attr-val → *Attr-list Conjunction*
Attr-list → *Attribute* | *Attribute*. *Attr-list*
Attribute → *identifier*
Coreference → **#***identifier*
Diff-list → **<! !>** | **<!** *Conjunction-list* **!>**
Conjunction-list → *Conjunction* | *Conjunction* **,** *Conjunction-list*
List → **<>** | **<** *Conjunction-list* **>** | **<** *Conjunction-list* **,** ... **>** |
　　　　　< *Conjunction-list* . *Conjunction* **>**

Strings are are indicated by " (e.g., "me") or ' (e.g., 'Kim).

Comments are allowed in LKB files, but only in between descriptions. Comments are either single lines beginning with ; or they are bracketed by #| |#.

The BNF for lexical entries is as follows:

Lexentry → *LexID Avm-def*.
LexID → *identifier*

The BNF for rules is essentially identical to that for lexical entries:

Ruleentry → *RuleID Avm-def*.
RuleID → *identifier*

4.5 Writing grammars in the LKB system

To fully understand LKB grammars, you need to know a little about the system's requirements as to the way grammars are organized. The main classes of object within the LKB system are:

Types and constraints
Lexical entries
Grammar rules
Lexical and morphological rules

Apart from types and their constraints, all the rest of the objects are generically referred to as *entries*. They are all TFSs which are associated with an identifier. They are distinguished from each other because of their function within the system, with respect to parsing, for example. Two more minor categories of entry are also used in the sample grammars:

Start symbol descriptions

Parse node descriptions

This section contains details of all the different types of entry we have used in the grammars we have seen so far. Lexical and morphological rules are left to the next chapter. Further details of the capabilities of the LKB system for dealing with larger grammars and ones which have a different approach to syntax are given in Chapter 8.[32]

The source file organization in the LKB is based on the assumption that a single file contains only a single class of object. For instance, mixing grammar rules and lexical entries in a single file is not allowed. The grammar engineer may choose to split the descriptions in a particular class over multiple files, for example to have different type files for the types used in rules as opposed to those used in lexical entries. This often enhances modularity. All the source files are read into the LKB under the control of a *script file*.

4.5.1 The script file

The following is the script file for grammar **g4diff**, which illustrates how the files of the different sorts of object are loaded into the LKB system:

```
(lkb-load-lisp (this-directory) "globals.lsp")
(lkb-load-lisp (this-directory) "user-fns.lsp")
(load-lkb-preferences (this-directory) "user-prefs.lsp")
(read-tdl-type-files-aux
    (list (lkb-pathname (this-directory) "types.tdl")))
(read-tdl-lex-file-aux
    (lkb-pathname (this-directory) "lexicon.tdl"))
(batch-check-lexicon)
(read-tdl-grammar-file-aux
    (lkb-pathname (this-directory) "rules.tdl"))
(read-tdl-start-file-aux
    (lkb-pathname (this-directory) "start.tdl"))
(read-tdl-parse-node-file-aux
    (lkb-pathname (this-directory) "parse-nodes.tdl"))
```

---

[32]There are further ways to use entries in the LKB which are not described in the current book. These include translation equivalences (sometimes referred to as bilexical entries) and rules for semantic conversion.

The script file is written in Lisp, although for convenience a number of functions have been defined to make writing a script file easier and it is not necessary to know Lisp in order to modify a script file. A short description of Lisp syntax is given in §4.5.2. The script file functions are discussed in detail in §8.2. Most of the time it is easiest to work from an existing script file.

The first three lines of the script read in three Lisp files which parameterize the system in various ways. These files are briefly discussed in §4.5.8. The first two files are loaded by `lkb-load-lisp`, which is an LKB function that takes a directory and a file name. The third file is for the user preferences. For convenience, the function `this-directory` is provided to refer to the directory in which the script file is located (`parent-directory` is also available to refer to the directory above the script file's directory). The lines after that read in the types with their local constraint descriptions and then the entries of the various classes. The type files have to be read in before the entries and the lexicon must be loaded before the auxiliary entries (i.e., the start symbols and the parse nodes). All these files are read in by functions which take pathnames (i.e., descriptions of file locations). The pathname is most easily constructed via the function `lkb-pathname` which takes a first argument which is a specification of the directory which contains the file and a second argument which is the actual file name.

4.5.2 *A lightning introduction to Lisp syntax

This section briefly lists a few things about Lisp that you may want to know to follow some of the details of the script files. If you are happy not understanding all the details, skip this for now, and come back to it if necessary when you want to write your own grammars.

1. All function calls have the syntax:

   `(<fn-name> <arg-list>)`

   i.e., a function name followed by zero or more arguments. For instance

   `(this-directory)`

   calls the function `this-directory` with zero arguments.

2. Function calls may provide arguments for other functions:

   `(lkb-load-lisp (this-directory) "user-fns.lsp")`

   the call to `(this-directory)` provides the first argument to `lkb-load-lisp`.

3. Strings are enclosed in "s. For instance, the second argument in

```
(lkb-load-lisp (this-directory) "user-fns.lsp")
```

is a string.

4. Generally, all arguments are evaluated: to specify a symbol, which should not be evaluated, it must be preceded by ' — for instance, in the following example from the file globals.lsp, 'string is a symbol.[33]

```
(def-lkb-parameter *string-type* 'string)
```

5. t (true) and nil (false) are the Lisp boolean values. However, anything which is not nil is taken as true if interpreted as a boolean.

6. Lists are enclosed in (): nil is also used to indicate the empty list. Lists of symbols can be quoted:

```
(def-lkb-parameter *orth-path* '(orth list first))
```

Lists may alternatively be constructed by the function list which takes any number of arguments. For instance:

```
(list (lkb-pathname (this-directory) "types.tdl"))
```

constructs a list containing a single element.

4.5.3 Loading the type system

Type files are read in by read-tdl-type-files-aux which takes a list of type file pathnames (a list of one element in the example above).

The type system is defined in one or more type files. It is necessary to have loaded a valid type system before anything in the LKB will work other than the grammar loading commands. The system loads the type files and then carries out a series of checks and expands the type constraint descriptions. With an error-free type hierarchy, you will see messages such as the following while the system loads the files:

```
Reading in type file types.tdl
Checking type hierarchy
Checking for unique greatest lower bounds
Expanding constraints
Making constraints well formed
Expanding defaults
Type file checked successfully
Computing display ordering
Reading in lexical entry file lexicon.tdl
Checking lexicon
```

---

[33]Lisp functions evaluate all their arguments, but def-lkb-parameter is defined as a macro, not a function: however, all the ordinary LKB user needs to know is that the second argument to def-lkb-parameter should be quoted.

```
Lexicon checked
Reading in rules file rules.tdl
Reading in psort file start.tdl
Reading in templates file parse-nodes.tdl
Grammar input complete
```

Once a type file is loaded successfully, a window showing the type hierarchy is displayed.

If the syntax of a description is incorrect or a type system condition is violated, then error messages appear. Error messages are described in detail in §7.1, but briefly, the checks are as follows:

Syntactic well-formedness If the syntax of the constraint specifications in the type file is not correct, according to the definition in §4.4.6, then syntax error messages will be generated: see §7.1.1.

Conditions on the type hierarchy If the conditions on the type hierarchy are violated, error messages generally appear as described in §7.1.2. The exception is the unique greatest lower bound condition. The LKB requires that every set of types has a unique greatest lower bound (*glb*) as described in §3.2. However, the system can fix this problem without user intervention by introducing new types. These types are given names such as **glbtype1** etc, and I will refer to them as *glbtypes*. They never have their own local constraint descriptions.

Some glbtypes are introduced for grammar **g8gap** which we looked at in Chapter 2. By default, the type hierarchy is displayed without glbtypes, but a hierarchy with glbtypes can be shown by choosing **View Type Hierarchy** and setting the `Show all types` checkbox. **g8gap** has a glbtype under **unary-rule** and another under **head-initial**. If you view the type definition of a type which has a glbtype as a parent, it will show the glbtype, but show the 'real' parents after it in parentheses. Apart from their user-interface properties, glbtypes behave exactly like user-defined types.

Constraints Once the type hierarchy is successfully computed, the constraint descriptions associated with types are checked, and inheritance and typing are performed to give expanded constraints on types (see §3.5). If this fails, error messages are generated as described in §7.1.3.

4.5.4 Loading lexical entries

For simple grammars, such as the example grammars in this book, lexicon files are read in by `read-tdl-lex-file-aux`. This takes a single pathname or a list of pathnames, if there are multiple lexicon files.

When lexicon files are loaded, they are generally only checked for syntactic correctness (as defined in §4.4.3) — entries are only fully expanded when they are needed during parsing or generation or because of a user request to view an entry. This is because large lexicons take up a large amount of memory. Thus when loading the lexicon, you may get syntactic errors similar to those discussed in 7.1.1 above, but the content errors will only be detected when you view a lexical entry or try to parse with it. However, the sample grammars contain a call in the script file to `batch-check-lexicon`. This is a function (with no arguments) which checks the lexicon during loading, so error messages appear at that point. This is convenient with small grammars.

The lexicon is stored in a temporary file when it is read in, not in main memory. This file is stored in the temporary directory that you were asked to construct in §2.1 — i.e., `/tmp` on Linux or Solaris, or `C: tmp` on Windows. The default filename is `templex`: see §9.3.1 for details.

Large lexicons may be cached to avoid the expense of rereading them each time a grammar is reloaded: see §8.8.

4.5.5   Loading grammar rules

Grammar rules are read in by `read-tdl-grammar-file-aux` which takes a single pathname. Grammar rules are always expanded at load time. If there is an error you will get error messages when the rules are loaded. Rules must expand out into TFSs which have identifiable paths for the mother and daughters of the rule. For the grammars in this book, the mother path is the empty path and the daughter paths are defined in terms of the ARGS feature. For example:

$$
\begin{bmatrix}
\textbf{binary-rule} \\
\text{ARGS} \begin{bmatrix} \text{FIRST } \textbf{sign} \\ \text{REST} \begin{bmatrix} \text{FIRST } \textbf{sign} \end{bmatrix} \end{bmatrix}
\end{bmatrix}
$$

Here the mother is given by the empty path, one daughter by the path ARGS.FIRST and another by the path ARGS.REST.FIRST. The mother path and a function which gives the daughters in the correct linear order must be specified as system parameters in the `globals` and `userfns` files (see §9.2.3 and §9.3).

4.5.6   Start structures

As discussed in 4.1, a valid parse is defined as a well-formed phrase that spans the entire input string and which also satisfies any *start structure* conditions that the grammar writer may impose. For instance, *Kim sleep*, with non-finite *sleep*, might be a valid phrase (because of sentences such as *Sandy prefers that Kim sleep*), but blocked from being a sentence because of a start structure condition that requires a finite main verb.

Start structure conditions are implemented in the LKB via a parameter *start-symbol* which points to specifications of TFSs. Like other grammar-specific parameters, *start-symbol* is set in the files that are loaded at the beginning of the script. If the value of *start-symbol* is nil, then any phrase which spans the input is accepted. Otherwise, the value of *start-symbol* may be a single identifier or a list of identifiers of feature structures: the TFS associated with a phrase must unify with at least one of the start symbol TFSs in order to be accepted as a valid parse. Although start structures are usually implemented with entries, the value of *start-symbol* may be a type name, in which case the start structure TFS is the constraint on the type.

The value of *start-symbol* may be altered interactively, via the **Options / Set options** menu command: if suitable start structures are defined this allows the user to switch easily between allowing only full sentences and allowing fragments. For instance, setting the start symbol to **phrase** in the example grammars will allow any fragments to be accepted as valid parses (e.g., parsing *the dog* results in a parse tree being displayed).

In the sample grammars, the start symbols are entries which are stored in a separate file. This is read in by read-tdl-start-file-aux which takes a single pathname. Start symbols may be viewed via **Lex Entry** on the **View** menu. The start symbols are not actually lexical entries, but to avoid having too many items on the **View** menu, **Lex entry** is used as a default for viewing 'other' entries too.

4.5.7 Parse tree node labels

Parse tree node labels are read in by read-tdl-parse-node-file-aux which takes a single pathname.

The concept of parse trees was discussed in §4.1.2: in this section, I go into a little more detail about the way the nodes get their labels. The parse node file allows TFSs which control the labelling of the nodes in the parse tree to be defined. I will refer to these structures as *node templates*: in effect, templates are pairings of node labels and TFSs. In very general terms, a parse tree node will be get the label associated with a template, if the TFS associated with the node is matched by the template's TFS. The format for parse node descriptions is identical to that of lexical entries. Parse nodes may be viewed via **Lex Entry** on the **View** menu.

The precise behaviour of the node labelling depends on the value of the parameter *simple-tree-display*. If *simple-tree-display* is set to **t**, as in the sample grammars for this book, then the templates consist of identifiers associated with single feature structures. If the template TFS subsumes the structure constructed for a node in a parse, the node is labelled with the identifier of the template. The first matching template is used. If no template matches, then the node is labelled with the type of its own TFS. A more complicated approach is used if *simple-tree-display* is set to nil, as is described in §8.14.

4.5.8 Ancillary files

These are the files loaded at the beginning of the script which parameterize the behaviour of the system.sers who are working with existing grammars should not need to know much about these files, so details are left to Chapter 8. All the ancillary files are loaded by lkb-load-lisp, which is an LKB function that takes a directory and a file name. There is an optional third argument, which if set to t indicates that the file need not be present.

Globals Default file name globals.lsp. Contains parameters that may be changed by the user. Details are given in Chapter 9.

User defined functions Default file name user-fns.lsp. There are a few functions which control parameters such as rule ordering which may differ in different grammars. These are in the user-fns file (written in Lisp). Details are given in Chapter 9.

User preferences Default file name user-prefs.lsp. This file is optional — it is loaded by a special command.

```
(load-lkb-preferences
    (this-directory) "user-prefs.lsp")
```

This file contains parameter settings for parameters which can be controlled interactively via **Set options** on the **Options** menu (see §6.1.9). Specifying this file in the script means that any preferences which are set in one session will be saved for a subsequent session. The user should not need to look at this file and should not edit it, since any changes may be overwritten. Details of the individual parameters are given in Chapter 9.

4.6 Summary

In this chapter, we followed up on the previous chapter by looking more precisely at how TFSs are used in parsing and generation, both abstractly and in an actual implementation. We also discussed derivations, parse trees and parse charts and introduced some aspects of underspecification. This chapter illustrated some more advanced techniques using TFSs, especially difference lists. It also went through the way that LKB grammars are written in considerable detail, covering the description language and the file organization. The exercises in this chapter were designed to help you explore parsing and to start modifying and expanding simple grammars.

5

More advanced grammars

The purpose of this chapter is to explain some of the details of more complex and linguistically interesting grammars. To start off with, we will look at a simple grammar which uses a more *lexicalist* style of grammar encoding and is thus closer to the grammar frameworks which TFS systems are most often used to encode, Head-Driven Phrase Structure Grammar (HPSG) in particular. This grammar still has a full form lexicon (i.e., all the inflectional variations are listed explicitly) but then as a next step we will look at the treatment of morphology and lexical rules in the LKB system.[34] The rest of the chapter briefly illustrates some of the principles of encoding semantics in TFSs and discusses some other grammar encoding issues, including long distance dependencies. By the end of the chapter, we will have finally worked up to the grammar that was used in the tour of the LKB in Chapter 2.

5.1 A lexicalist grammar

In the previous chapter, we looked at grammars which had a very obvious resemblance to conventional CFGs. We used unification to provide a better account of agreement than is possible with a standard CFG, but the grammars still used a very conventional style of grammar rules. Lexicalist grammar frameworks move much more information away from the grammar rules and into the lexicon (or into types encoding lexical generalizations).

To see why lexicalist grammars are often considered to be attractive, let's consider what happens when we enlarge our earlier grammar. For instance, suppose we want to extend it to capture verbs other than intransitive ones. We could write rules such as the following:

---

[34]The treatment of lexical and morphological rules in unification-based systems is not at all standardized, so most of this discussion is LKB-specific.

```
;;; VP -> Vsimple-trans NP

trans_v_rule := phrase &
[ CATEG vp,
  ARGS [ FIRST [ CATEG v,
                 VTYPE simple-trans ],
         REST [ FIRST [ CATEG np ],
                REST *null* ]]].

;;; VP -> Vvping VPing

trans_vping_rule := phrase &
[ CATEG vp,
  ARGS [ FIRST [ CATEG v,
                 VTYPE vping ],
         REST [ FIRST [ CATEG vp ],
                REST *null* ]]].

;;; VP -> Vpp PP

trans_pp_rule := phrase &
[ CATEG vp,
  ARGS [ FIRST [ CATEG v,
                 VTYPE pp ],
         REST [ FIRST [ CATEG pp ],
                REST *null* ]]].
```

Each rule specifies the category of its mother and its daughters. I have ignored agreement for simplicity, but it could be put in as before. The lines ;;; VP -> Vpp PP, and so on, are simply comments to show an equivalent conventional CFG.

One consequence of following this approach is that we end up with a large number of VP rules, each of which is capturing one subcategorization possibility. The verbs themselves have to be lexically marked in order to make sure that only the correct rules apply, indicated here by the feature VTYPE. There has to be a correspondence between these types and the rules, so we are, in some sense, duplicating information between the grammar and the lexicon.

Unification based grammars, however, allow an alternative encoding style which greatly simplifies the rule system (though it somewhat complicates lexical signs). In this approach, there is only one rule for combining a verb with a complement to form a VP. In fact, we can go further, because this single rule will also allow for the combination of a noun or an adjective with a complement. The trick is for lexical entries to specify the complements that they can take as a list. For example, in this revised approach, which is illustrated in grammar g5lex, the entry for *chased* is as follows:

```
chased := word &
[ ORTH "chased",
  HEAD verb,
  SPR < phrase &
       [ HEAD noun,
         SPR <>] >,
  COMPS < phrase &
          [HEAD noun,
           SPR <>] > ].
```

This entry has a feature COMPS, which takes a list of the complements to the verb: in this case there is a single NP complement. It also has a feature SPR which specifies the subject (again as an NP) but I will concentrate on the COMPS list for the time being.

The rule for combining a word with a single complement defined is as follows:

```
head-complement-rule-1 := phrase &
[ HEAD #head,
  SPR #spr,
  COMPS < >,
  ARGS < word &
         [ HEAD #head,
           SPR #spr,
           COMPS < #nonhddtr > ],
         #nonhddtr >  ].
```

This is a binary rule: it has two elements in its ARGS list. However, all that is said about the second daughter is that it is coindexed with the single element of the COMPS list of the first daughter (as specified by the reentrancy tag #nonhddtr). If the lexical entry for *chased* is the TFS that instantiates the first daughter position, a constraint arises on the second daughter that it has to be a **phrase** with HEAD **noun**. The first daughter is said to be the *head daughter* of the rule: this is intended to correspond to the notion of headedness that is traditional in linguistics. The phrase as a whole has an empty COMPS list: the application of a rule in this way is sometimes said to *discharge* the complement requirement

for the verb.[35] However, the head daughter's value for the SPR is shared
with the phrase as a whole (**#spr**), so in the case of *chased this dog*,
the phrase has an undischarged SPR slot. The rule also specifies that
the HEAD of the phrase and the head daughter are shared(**#head**). The
structure for the phrase *chased this dog* is shown below — I have omitted
the ARGS of the daughters.

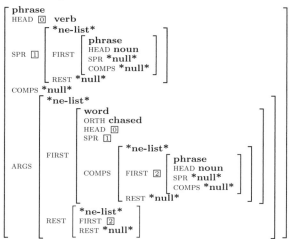

Compare the `head-complement-rule-1` to the more traditional CFG-
style rules shown earlier: given suitable lexical entries, this single rule
would stand in for all of those rules.

A full grammar based on this encoding style is given in `g5lex`. Because I want to make grammar `g5lex` small and make the comparison
with grammar `g1cfg` as clear as possible, `g5lex` does not include agreement or difference list encoding of orthography. It also does not make
maximal use of the type hierarchy to encode generalizations, so the lexicon is somewhat verbose and would be highly repetitive if we had more
than a few lexical entries: grammars we will look at later on will avoid
such redundancy.

The particular feature architecture in grammar `g5lex` is roughly
based on a small fragment of Sag and Wasow (1999), which should be
referred to for a detailed explanation of the linguistic motivation. A
grammar that corresponds to a fairly full implementation of Sag and
Wasow is distributed with the LKB as the `textbook` grammar. However, it should be possible to follow this chapter without having read

---

[35]This terminology is intuitive and useful but requires a health warning — it sounds
as though processing is necessarily bottom-up and of course the rule is neutral about
processing order, like any other rule in a unification-based grammar.

Sag and Wasow (1999), especially if you spend some time experimenting with the sample grammars. Our aim here is to look at how this grammar exploits TFS and unification to move away from conventional CFG style grammars. The main differences between this grammar and our initial grammar g1cfg are:

1. The main objects are of type **sign** rather than **syn-struc**.

2. All **sign**s have a COMPS which takes a list value. Different **words** have different values for COMPS. For instance, a simple transitive verb, such as *chased*, specifies that its COMPS list has a single element which is a **phrase** with a HEAD value of **noun** and SPR value empty (i.e., an NP, see below). In contrast, an intransitive verb, such as *barks*, has an empty COMPS list. The head complement rules (head-complement-rule-0, head-complement-rule-1 and head-complement-rule-2) are responsible for combining a word (the *head daughter* of the rule) with its complements.[36] In the case of chased, the head-complement-rule-1 will apply when *chased* is followed by an NP.

   Grammar g5lex has a rule which applies to elements with a null complements list: head-complement-rule-0. This is a unary rule: i.e., it only has one daughter.

   The head daughter in a head complement rule is always of type **word**, the mother is of type **phrase**. Apart from the head in the head complement rules, all the other daughters of rules in this grammar are of type **phrase**.

   The head-complement-rule-0 just converts a **word** to a **phrase**: it will apply to *barks*, for instance. Unary rules of this sort are sometimes referred to as *pumping rules*: the rule is said to *pump* a **word** to a **phrase**.[37]

3. Instead of g1cfg's CATEG feature, g5lex uses a feature HEAD which takes a value **pos**. In this grammar, this is an atomic type, with possible values **noun**, **verb** and **det**. Categories such as VP are

---

[36]Readers familiar with Sag and Wasow will note that they assume that only a single head complement rule is required, but allowing rules to have an arbitrary number of daughters has unpleasant formal and computational ramifications. Most linguists assume that there is an upper bound on the number of complements that a word can have, so we can get the right effect by having multiple head complement rules. We will see in §5.3.2 that we can capture generalizations that apply to all the rules by using the type hierarchy. An alternative to having one rule for each number of complements is to adopt a binary branching approach, where each rule application combines one complement with the head. A grammar that adopts this approach is briefly discussed later in this chapter.

[37]The health warning about procedural terminology in the earlier footnote applies here too.

implicit: we can use the term VP to talk about a sign which has HEAD equal to **verb**, an empty COMPS list and a non-empty SPR list, but it is just a convenient abbreviation for this bundle of properties.

4. All rules in `g5lex` equate the feature HEAD of the mother with the HEAD of the head daughter. Head daughters are not explicitly distinguished in this grammar, but we could do this if we wanted to by defining a feature HEAD-DTR which had a value which was coindexed with one of the elements of the ARGS list.

5. All **signs** have a SPR which takes a list value. SPR is either a single element list or an empty list. SPR is used both for the subject of a verb and for the determiner of a noun. The **head-specifier-rule** combines a specifier with a head. Note that while in the head complement rules, the head is always the first daughter (since we are only dealing with English), in the head specifier rule, the head follows the specifier.

5.1.1 Exercises

Before trying these exercises, make sure you understand how the grammar in `g5lex` works. Note that `g5/test.items` gives a list of test sentences. Some of these sentences are ungrammatical, but will nevertheless parse: i.e., the grammar *overgenerates*. Conversely, some of the sentences are grammatical but will not parse, so the grammar also *undergenerates*. The objective when grammar engineering is to increase coverage while avoiding overgeneration: a set of test sentences is essential to make sure this objective is being met as the grammar is changed.

Some of these exercises in this section are quite hard: it is likely to take several hours to complete the whole set. You will probably need to look at the reference chapter on error messages and debugging (Chapter 7).

1. The rule that is needed for ditransitive verbs is in the grammar, but there are no lexical entries that utilize it. Add an entry for the ditransitive verb *gave* taking two NP complements to `g5lex`, so you can parse *the dog gave the cat the aardvark*. The easiest way to do this is to copy the entry for *chased*, but change the orthography and add another element to the COMPS list. Make sure you cannot parse **the dog gave the cat*. Add these sentences to `test.items`.

2. Add a new type for prepositions, an entry for *to* and another entry for *gave*, so that you can parse *the dog gave the aardvark to the cat*. The type **prep** can be a new subtype of the type **pos**. The

lexical entry for *to* will have a value for HEAD of **prep**. It should take a single NP complement, like *chased*, but unlike *chased*, have an empty SPR. The new lexical entry for *gave* will specify that the second member of its COMPS list is a PP: i.e., something with HEAD **prep** and COMPS empty. (Don't bother about making sure the preposition in the PP is restricted to *to* for now.) As before, add appropriate sentences to `test.items`.

3. Modify your grammar to enforce the constraint that the value of SPR is a list of no more than one element. Right now, a lexical entry with a SPR list longer than one will not result in any parses, but it is better to detect such errors at the point when the lexicon is loaded. You could enforce this more specific constraint by defining a subtype of list called **max-one-list**. This type should be a supertype of a new type that corresponds to a list containing exactly one element. It should also be a supertype of the empty list. Specify that SPR should take **max-one-list** rather than ***list***. Check the constraint by adding a buggy entry to the lexicon with a two element SPR list and ensuring that an error is signalled. Make sure that the test sentences in `test.items` still behave as before.

4. Augment your new grammar so that orthography is built up by difference lists in the manner of `g4diff`. This involves several changes:

   (a) Add the difference list type to `types.tdl` and modify the value of ORTH so it takes a difference list.

   (b) Change the lexical entries so the orthography is in the right place, as in `g4diff`.

   (c) Modify the grammar rules to build up the orthography: note that we now have unary, binary and ternary rules, so there should be different patterns of difference list construction for each case.

   (d) Modify the value of `*orth-path*` in `globals.lsp` and add the difference list values as in `g4diff`.

   Make sure that the test sentences in `test.items` still all parse. Ensure they now have the correct values for ORTH.

5. Augment your new grammar so that it encodes agreement. Make NUMAGR be an appropriate feature of the type **pos** with values as in the grammar `g2agr`. Take advantage of the SPR so that coindexation between agreement values is enforced in the lexicon rather than in the rules. For instance, the agreement value for a noun can be coindexed with that of its determiner via the noun's SPR feature. Make sure that the test sentences in `test.items` that

should parse still do so, but that you have excluded the ones with incorrect agreement.

6. Modify your last grammar so that only nouns and determiners carry the NUMAGR feature. To do this, you can add a new type, **nominal**, which is a subtype of **pos** and a supertype of **det** and **noun**. Verbs do not themselves have NUMAGR: they simply stipulate the NUMAGR of their SPR as appropriate.

5.1.2 Answers

Grammars `an6`, `an7` and `an8` provide solutions to the various steps with comments: a grammar containing all the required modifications is in `an8`.

5.1.3 *Categorial grammars

The HPSG style of grammar encoding is simply one possibility for lexicalist grammars. An alternative approach, which is even more lexicalist, is *categorial grammar* (for general details of categorial grammar, see Wood, 1993). A sample grammar, roughly equivalent to `g5lex`, but using a categorial grammar style of encoding, is given in `catgram`. Categorial grammar is based on a very uniform approach, where rule application involves combining a functor with an argument. The functor is analogous to the head daughter in the rules we looked at in `g5lex` and the argument is analogous to the complement or specifier: however, no distinction is made between different classes of argument. In the grammar in `catgram` there are just two binary rules: `forward-application` for a rule where the functor comes first and the argument second, and `backward-application` where the functor is second. The directionality that applies for each argument is encoded lexically, via the feature DIRECTION. Although realistic categorial grammars generally require a few more rules besides forward and backward application (for reasons I won't go into here), they are all highly abstract and the number of rules is still considerably less than a full HPSG (e.g., the LinGO ERG has around 40 rules). But the differences between the approaches are really rather unimportant as far as the LKB system encoding is concerned, and in fact the successively more complex grammars that we will look at in this chapter could equally well have been based on categorial grammar (see the exercise in §5.4.6).

5.2 Lexical and morphological rules

One obvious problem with the grammars we have seen so far is the need to explicitly list all the inflectional variants: e.g., *dog, dogs*; *chase, chases, chased*. Clearly we need some way of encoding inflection. There is also some redundancy in requiring two distinct entries for *give* (e.g., for *gave the dog the aardvark* and *gave the aardvark to the dog*). Comparable dual forms exist for a considerable number of ditransitive verbs: e.g., *hand, pass, fax*: this is usually referred to as the *dative alternation*. The basic mechanism for dealing with both these causes of redundancy in the LKB is a variety of unary rule: *morphological rules* are used if affixation is needed and *lexical rules* if no affix is involved. Lexical rules and morphological rules in the LKB differ only in that the morphological rules have additional annotation indicating the affixation patterns. So, where there is no possibility of confusion, I will sometimes use the term *lexical rule* to apply generically to rules with and without affixation: the LKB menu commands adopt this usage.

Morphological rules require a separate component to describe the spelling effects, which I discuss in detail below in §5.2.1. In the LKB, lexical rules without morphological effect differ from unary grammar rules only in that they may apply before morphological rules. In general, lexical and morphological rules can be applied in any order that is licensed by the affixation: thus lexical rules can be interleaved with morphological rules. In principle, lexical rules may also be applied to phrases, so all conditions on lexical rule application must be explicitly encoded, e.g., by creating a type **lexical-rule** for all lexical rules, with a daughter slot that is a **word**s rather than a **phrase**.

The grammar in g6morph uses morphological rules to encode inflection. Although this is based on g5lex (in fact on an8), several changes have been made. There is an additional file, inflr.tdl, that contains the inflectional rules and the script file includes the following line in order to load it:

```
(read-morph-file-aux
  (lkb-pathname (this-directory) "inflr.tdl"))
```

The main difference in the underlying architecture of the TFSs is that a type **lexeme** has been introduced for uninfected forms: inflectional rules map between **lexeme**s and **word**s. All the entries in the lexicon are of type **lexeme**.[38]

For instance, *chases* can be constructed by applying the 3sg-v_irule to the lexeme *chase*. The rule 3sg-v_irule is specified as follows:

```
3sg-v_irule :=
%suffix (!s !ss) (!ss !ssses) (ss sses)
  sing-verb.
```

---

[38] For more details of the linguistic motivation for this distinction, see Sag and Wasow (1999:175).

The information following the % is the affixation pattern, to be discussed later. The TFS is simply specified to be of the type **sing-verb**. This rule is shown in expanded form as an AVM below:

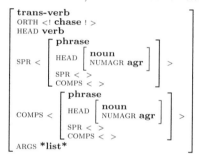

$$
\begin{bmatrix}
\textbf{sing-verb} \\
\text{ORTH} \;\boxed{0}\; \oplus +\textbf{s} \\
\text{HEAD} \;\boxed{1}\; \textbf{verb} \\
\text{SPR} \;\boxed{2}\; < \begin{bmatrix} \textbf{phrase} \\ \text{HEAD} \begin{bmatrix} \textbf{noun} \\ \text{NUMAGR } \textbf{sg} \end{bmatrix} \\ \text{SPR} < > \\ \text{COMPS} < > \end{bmatrix} > \\
\text{COMPS} \;\boxed{3}\; \\
\text{ARGS} < \begin{bmatrix} \textbf{verb-lxm} \\ \text{ORTH} \;\boxed{0}\; \\ \text{HEAD} \;\boxed{1}\; \\ \text{SPR} \;\boxed{2}\; \\ \text{COMPS} \;\boxed{3}\; \end{bmatrix} >
\end{bmatrix}
$$

I have used the abbreviation $\oplus +s$ in the ORTH to informally indicate the affixation. I have also abbreviated lists by using < >.

This rule applies to **lexeme**s, such as that for *chase*:

$$
\begin{bmatrix}
\textbf{trans-verb} \\
\text{ORTH} <! \textbf{ chase } ! > \\
\text{HEAD} \textbf{ verb} \\
\text{SPR} < \begin{bmatrix} \textbf{phrase} \\ \text{HEAD} \begin{bmatrix} \textbf{noun} \\ \text{NUMAGR } \textbf{agr} \end{bmatrix} \\ \text{SPR} < > \\ \text{COMPS} < > \end{bmatrix} > \\
\text{COMPS} < \begin{bmatrix} \textbf{phrase} \\ \text{HEAD} \begin{bmatrix} \textbf{noun} \\ \text{NUMAGR } \textbf{agr} \end{bmatrix} \\ \text{SPR} < > \\ \text{COMPS} < > \end{bmatrix} > \\
\text{ARGS } \textbf{*list*}
\end{bmatrix}
$$

I have indicated the difference list encoding of the orthography by using the <! !> notation.

If the lexeme for *chase* instantiates the daughter slot of 3sg-v_irule, the result is the TFS for the **word** *chases*, as follows:

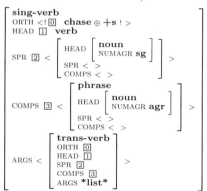

$$
\begin{bmatrix}
\textbf{sing-verb} \\
\text{ORTH} <! \boxed{0} \textbf{ chase } \oplus +\textbf{s} ! > \\
\text{HEAD} \;\boxed{1}\; \textbf{verb} \\
\text{SPR} \;\boxed{2}\; < \begin{bmatrix} \text{HEAD} \begin{bmatrix} \textbf{noun} \\ \text{NUMAGR } \textbf{sg} \end{bmatrix} \\ \text{SPR} < > \\ \text{COMPS} < > \end{bmatrix} > \\
\text{COMPS} \;\boxed{3}\; < \begin{bmatrix} \textbf{phrase} \\ \text{HEAD} \begin{bmatrix} \textbf{noun} \\ \text{NUMAGR } \textbf{agr} \end{bmatrix} \\ \text{SPR} < > \\ \text{COMPS} < > \end{bmatrix} > \\
\text{ARGS} < \begin{bmatrix} \textbf{trans-verb} \\ \text{ORTH} \;\boxed{0}\; \\ \text{HEAD} \;\boxed{1}\; \\ \text{SPR} \;\boxed{2}\; \\ \text{COMPS} \;\boxed{3}\; \\ \text{ARGS } \textbf{*list*} \end{bmatrix} >
\end{bmatrix}
$$

It should be clear that the 3sg-v_irule rule is operating just like a (unary) grammar rule in that the **lexeme** for *chase* can instantiate the single element of the ARGS list of the rule to give the **word**. Coindexations relate the various parts of the **lexeme** and the **word** structures: the effect of singular agreement is achieved by adding information to the SPR of the inflected form which enforces singular agreement on the subject noun phrase.

The grammar rules in g6morph are essentially unchanged from grammar g5lex (although the type hierarchy has been used to simplify their specification), hence they cannot apply to a **lexeme**. The approach of making every lexical entry be a **lexeme** follows Sag and Wasow (1999): it means that even words like *the* have a rule applying to them to pump them to type **word**. This rule is const-pump in inflr.tdl: rules without affixation, like const-pump, can be in the same file as rules with affixation as long as read-morph-file-aux is used to load the file.

Grammar g6morph makes much more extensive use of the type system to encode generalizations than does g5lex. For instance, recall that in g5lex there was a very verbose specification of *chase*, repeated below for convenience:

```
chased := word &
[ ORTH "chased",
  HEAD verb,
  SPR < phrase &
      [ HEAD noun,
        SPR <>] >,
  COMPS < phrase &
        [HEAD noun,
         SPR <>] > ].
```

The new entry for *chase* is simply:

```
chase := trans-verb &
[ ORTH.LIST.FIRST "chase" ].
```

We have achieved this simplification in the lexical entry via the type system, which specifies a type **trans-verb** that inherits from **verb-lxm**.

```
verb-lxm := lexeme &
[ HEAD verb,
  SPR < phrase & [HEAD noun, SPR <>] > ].
```

```
trans-verb := verb-lxm &
[ COMPS < phrase & [HEAD noun, SPR <>] > ].
```

Grammar rules have also been somewhat simplified compared to g5lex by making use of type constraints. As we saw above, the information in the inflectional rule 3sg-v_irule was all encoded in the type **sing-verb**. We discuss the use of type constraints to encode generalizations further in §5.3.2.

Grammar g6morph also has a **lexeme** to **lexeme** lexical rule, dative-shift-lrule, which encodes the dative alternation. This is shown below:

$$
\begin{bmatrix}
\textbf{ditranspp-verb} \\
\text{ORTH } \boxed{0} \\
\text{ARGS} < \begin{bmatrix} \textbf{ditransnp-verb} \\ \text{ORTH } \boxed{0} \end{bmatrix} >
\end{bmatrix}
$$

This rule is very simple, because it just states that its daughter is a **ditransnp-verb** while the lexeme as a whole is a **ditranspp-verb** shares its orthography with its daughter. The rule is in lrules.tdl which is loaded by the following line in the script file:

```
(read-tdl-lex-rule-file-aux
  (lkb-pathname (this-directory) "lrules.tdl"))
```

There is no derivational morphology in this grammar, but if there were, it could be encoded by **lexeme** to **lexeme** rules as well. But remember that the **lexeme/word/phrase** distinction is in no sense built into the LKB and it is just one possible way of distinguishing between classes of rules in the grammar.

With lexical and morphological rules, there are issues of *productivity*: it may be difficult to specify a natural class of words to which a lexical alternation or a derivational rule applies. For instance, the dative alternation applies to most verbs that can be construed as involving change of possession, but there are a few exceptions, such as *tip*:

I tipped Kim five pounds.
*I tipped five pounds to Kim.

Productivity is a complex and interesting topic, but we will not discuss mechanisms for restricting the application of lexical and morphological rules here, other than the type system. We can prevent the dative-shift-lrule applying to *tip*, for instance, by making a type distinction between *tip* and the normal ditransitive verbs. If we specify that normal verbs are **ditrans-np-normal**, while *tip* is of type **ditrans-np-no-dative**, and that the daughter of dative-shift-lrule is of type **ditrans-np-normal**, then the rule will not apply to *tip*.

5.2.1 Morphological rules

Morphological rules in the LKB system are unary rules with added orthographic information describing affixation. The underlying assumption is that affixes are associated with specific rules. A simple string unification approach is used for encoding the affixation patterns.[39] Affixation and any associated spelling changes is indicated by means of information which augments the standard unary rules. This information is supplied by a line starting with % that comes immediately after the := that follows the identifier in the rule, but before the remainder of the rule description. As examples, consider the 3sg-v_irule, repeated from above, and also the past-v_irule:

```
3sg-v_irule :=
%suffix (!s s) (!ss !ssses) (ss sses)
sing-verb.
```

```
past-v_irule :=
%suffix (* ed) (!ty !tied) (e ed) (!t!v!c !t!v!c!ced) (give gave)
past-verb.
```

The information introduced by % describes the spelling changes — the rest of the rule is just described in the standard TDL syntax, though in these examples all there is is a type name. Lexical rules which don't affect affixation may be in a file with morphological rules: the former are simply ignored by the component of the grammar loader that handles spelling changes.

The spelling rule introduced by % has two parts — the first is simply either **prefix** or **suffix** and the second contain a set of subrules in the form of matched simple partial regular expressions. The first of each pair of expressions must match letters in the word stem, either at the beginning of the stem (for prefixes) or at the end (for suffixes, as illustrated here). For example, the subrule (e ed) will match the stem *chase*. The second expression in the subrule describes the letters in the affixed form of the word. Thus the stem *chase* corresponds to the affixed form *chased*.

The asterisk * and the symbols introduced by ! are special characters in spelling rules. The asterisk represents a null character, so in the entry 3sg-v_irule, (* ed) means that *ed* is added at the word-final position. Subrules are checked right to left when generating, so the * is usually used in the leftmost subrule to catch all the cases which do not match any of the other subrules. For instance, none of the other subrules in 3sg-v_irule match the stem/form pairing *bark/barked*. The character ! is used to denote special macro symbols which correspond to sets of

---

[39]This was implemented by Bernie Jones and the following description is adapted from text written by him. This part of the LKB is really just a placeholder until a more complete approach can be integrated: it works adequately for English and other languages with simple morphology but a more sophisticated system is required for many other languages.

letters. These sets are defined at the beginning of the file that contains the morphological rules. The set used in `inflr.tdl` is as follows:

```
%(letter-set (!c bdfglmnprstz))
%(letter-set (!s abcdefghijklmnopqrtuvwxyz))
%(letter-set (!a abcdefghijklmnopqrstuvwxyz))
%(letter-set (!t bcdfghjklmnpqrstvwxz))
%(letter-set (!v aeiou))
```

The interpretation of these symbols is that they match any one of the characters in the set, but the matching character must be the same throughout the subrule. For instance, (!t!v!c !t!v!c!ced) will match (*pot potted*), but would not match (*pot podded*).

The `3sg-v_irule` rule has a leftmost subrule that uses !s, which according to the letter set definition stands for any character other than s. This prevents the leftmost subrule applying to a word ending in 's'. This is better than using the *, which allows pairs like (*toss/tosss*) — in fact the version of `past-v_irule` shown above has the disadvantage that *chase/chaseed* is an acceptable pairing.

Morphology is reversible, and can thus be used for parsing and generation. However, unlike most other parts of the LKB system it is asymmetric, in that it accepts input that it won't generate. When analysing, the system accepts any form that matches a pattern. When generating, the system will only produce one form, with priority given to the rightmost matching subrule. So for instance, given the rule for `3sg-v_irule` above, the analyser will accept *chaseed* as well as *chased* but the generator will only produce *chased*.

The subrule system can be used to allow for irregular affixation, as illustrated above by the subrule (`give gave`). Because this is just interpreted as matching the characters at the end of the string, it will match the stem *give* and also *forgive*. An alternative method for specifying irregular inflectional morphology is discussed in §8.4.

5.2.2 *BNF for morphology

For completeness, I give the BNF for the morphological specifications in this section. The BNF for the letter sets is as follows:

Letterset → %(**letter-set** (*Macro letters*))
Macro → !*letter*

where *letter* corresponds to any character, *letters* to one or more characters and **letter-set** is a literal: i.e., 'letter-set' must appear in the file.

The rules themselves have the following BNF:

Mruleentry → *RuleID Mgraph-spec Avm-def.*

Mgraph-spec → **%prefix** *SPair-list* | **%suffix** *SPair-list*

SPair-list → *SPair* | *SPair SPair-list*

SPair → (* *Char-list*) | (*Char-list Char-list*)

`Char-list` -> *letter* | *Macro* | *letter Char-list* | *Macro Char-list*

Avm-def is as defined in §4.4.6.

5.2.3 Using lexical and morphological rules

There is considerable flexibility in the way rules are applied, since as far as possible the LKB does not build in any unnecessary assumptions about processing. As far as the parser and generator are concerned, the primary distinction is between rules that affect spelling changes (i.e., morphological rules) and those that don't. The system makes the assumption that all morphological processing happens at the level of an individual word (although see §8.5). Thus, when parsing, all affixation operations apply before grammar rules.[40] This means that it is the grammar writer's responsibility to stop rules applying in unwanted ways. On the whole, this is done via the type system, as exemplified above in §5.2. However there are two situations for which this mechanism is insufficient, or undesirably clunky. The first case is the marking of rules which have associated spelling effects. This is necessary to prevent these rules being applied when they are not licensed by the spelling. This is done by means of a user-definable function, `spelling-change-rule-p`, which specifies whether or not a rule affects spelling (see §9.3). In the version of the file `userfns` supplied with grammar `g6morph`, this checks the rule to see if both the following conditions are true:

1. the type of the rule is **word** or **lexeme** or a subtype of either of those types
2. the value of the orthography path is not the same in the mother and daughter: i.e., given that the value of `*orth-path*` is `(orth list first)`, whether there is a coindexation between the paths ARGS.FIRST.ORTH.LIST.FIRST and ORTH.LIST.FIRST.

The other case in which a rule should not be applied by the parser is when it is a redundancy rule: i.e., a rule which is not intended to be applied productively, but only used in lexical entries. However, I will not discuss this situation in this book.

---

[40]There are some examples in English morphology which may make this assumption problematic. For example, there is a suffix *-ed*, which forms adjectives such as *broken-hearted*, and which could be described as applying to an adjective noun phrase to form an adjective. Note that the adjective is obligatory (**a* hearted lover*), the result is sometimes but not always hyphenated, and that the process is productive (e.g., *rattan chaired terrace*).

5.2.4 Loading lexical and morphological rules

Lexical rules are read in by `read-tdl-lex-rule-file-aux`, which takes a single pathname. For instance:

```
(read-tdl-lex-rule-file-aux
    (lkb-pathname (this-directory) "lrules.tdl"))
```

Morphological rules must be read in by `read-morph-file-aux`, which also takes a single pathname.

```
(read-morph-file-aux
    (lkb-pathname (this-directory) "inflr.tdl"))
```

`read-morph-file-aux` accepts the spelling component of the rules, although it is not an error to include rules with no spelling component in a file that it reads. Lexical and morphological rules are expanded as they are read in, just like grammar rules. The conditions on mother and daughter paths are the same as grammar rules.

5.2.5 Lexical and morphological rules in parse trees and parse charts

Although lexical rules are really just like grammar rules in the LKB system, they are normally omitted from parse trees in linguistics. They can also clutter up parse charts. The LKB system therefore has parameters which can be set to control whether or not morphological and lexical rules are displayed: `*show-morphology*` and `*show-lex-rules*` respectively. See §9.1.2 for details. Both types of rule are shown by default. In parse trees, any rules which involve affixation are shown as branches under the representation of the input word. For instance, Figure 13 shows an LKB parse tree for *the dogs gave the cat to the aardvarks* with both `*show-morphology*` and `*show-lex-rules*` set to t.

5.3 Exploiting the type system in grammar encoding

The type system in the LKB serves several different functions in grammar encoding. In this section, I briefly discuss some of these functions, in order to illustrate various grammar encoding principles. In doing this, I will also review some of the features of the grammars we have seen so far.

5.3.1 Types and debugging

One function of the type system is analogous to the use of types in many programming languages: it allows some classes of error to be detected at load time. The type system allows the automatic detection of many classes of error in grammar writing: typographic errors in type and feature names, misplaced features, inheritance clashes and so on. If you

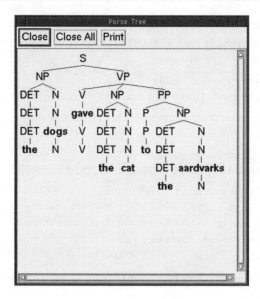

FIGURE 13 Parse tree illustrating the morphological and lexical rule display

have tried any substantial amount of grammar writing, you will know that getting the grammar to a state where it is accepted by the LKB can be a somewhat time-consuming and frustrating process, but it is far easier to do this than it is to debug grammars in untyped languages when these errors only appear when you try to parse a sentence! The LKB load time error messages provide a reasonably straightforward indication of where the problem is, while problems that arise when parsing a sentence require many more structures to be examined. Debugging tools and error messages are described in Chapter 7.

5.3.2 Types for inheritance

The most important use of types is to capture generalizations in the lexicon and in lexical rules and grammar rules. Without some form of inheritance, the lexicon would be extremely bulky in lexicalist approaches to grammar. Rules can also benefit from the use of inheritance, although with a lexicalist approach, the need is somewhat less. The grammar in **g6morph** begins to show the power of types for encoding generalizations. If you examine the types under **lexeme**, you will see how information is distributed so that repetition is avoided. The ideal situation is sometimes expressed as 'one fact, one place'. For instance, the type **noun-lxm** encodes the constraint that a noun must agree with its determiner: this is not repeated elsewhere in the grammar.

Generalization is extremely important for grammar engineering, since it allows a degree of localization of any changes: we could change the definition of **noun-lxm** without necessarily having to change any other type or any lexical entry. This is analogous to the use of inheritance in object-oriented programming. Encapsulating generalizations helps maintainability and increases the reusability of parts of the grammar. It also makes the grammar much easier to read, partly simply because the bulkiness of the files is reduced, but also because it helps the grammar engineer to ignore detail that is currently irrelevant. In principle, if a well-designed type system is developed by one grammar engineer, it can be used by another who doesn't necessarily know all the details, but just uses the types. For instance, the syntax and semantics might be developed by different people.

The rule hierarchy in grammar **g6morph** shows some examples of multiple inheritance. For instance, the type **head-initial** encodes the idea of a head daughter as the first daughter in a phrase, then this is used in multiple inheritance in conjunction with the types for the different arities of rule, as shown below:

```
head-initial := phrase &
[ HEAD #head,
  SEM [ INDEX #index ],
  ARGS < [ HEAD #head,
           SEM [ INDEX #index ] ], ... > ].

unary-head-initial := unary-rule & head-initial.

binary-head-initial := binary-rule & head-initial.

ternary-head-initial := ternary-rule & head-initial.
```

Multiple inheritance tends to get used more often in larger grammars. For instance, the concept of different complementation patterns is relevant for adjectives as well as verbs. Most adjectives are intransitive, but we also find ones that take complements, such as *easy*:

Kim is easy to please.

If we want to capture this, we can split the concept of complementation pattern away from part of speech, and cross-classify both verbs and adjectives with types such as **intransitive, inf-complement** and so on. It is worth noting though that a lot of cross-classification gets done in typed feature structure grammars without explicit multiple inheritance, because we can have different hierarchies for types that occur in different places in the feature structure of the sign. For instance, the type **pos** is the value for the feature HEAD which itself has subtypes.

Despite all the advantages of the use of inheritance, there can be situations when achieving the ideal distribution of information results in a very complicated type hierarchy. Some of these situations can be alleviated by the use of defaults, but although default inheritance is supported in the LKB, it is not discussed in detail in this book. In any case, more intuitive encodings may sometimes result if the hierarchy is kept simple, even if at the cost of some duplication of information.

5.3.3 Encoding information with types versus features

You may have started to wonder what the criteria are for encoding information with types or with features. In some cases, they are more or less interchangeable, and the choice made by a particular grammar writer may be essentially a matter of aesthetics. But in a lot of cases there is a distinct formal difference. I will illustrate this by going back to the agreement example.

The reason that the simple agreement example in the previous chapters worked out more nicely with TFSs than with the standard CFG was because the use of features to encode agreement allowed the specification of reentrancy between parts of the structure. If we had used subtypes of category instead, there would have been little advantage over the simple CFG version, because we could not have encoded the fact that we want the structures to match on agreement, but not on nouniness or verbiness, for instance. So features are used when we need to share information via reentrancy.

In the agreement example, we only considered number agreement. Of course, person is also relevant (and gender in many languages, but we'll ignore this here). One way of encoding this is to use a feature AGR instead of NUMAGR. AGR could have a value of type **pernum** with NUM and PER as appropriate features for **pernum**. For instance:

$$\left[\text{AGR} \quad \begin{bmatrix} \textbf{pernum} \\ \text{PER } \textbf{3rd} \\ \text{NUM } \textbf{sg} \end{bmatrix} \right]$$

We could then stipulate the reentrancy between the value of AGR on verbs and their subjects for instance, and ensure that they are consistent with respect to both number and person.

However, although this encoding has frequently been used, it may not be the most perspicuous way to treat agreement in English. The problem is that for most English verbs, the distinction in morphological form is between third person singular forms and everything else. While we can represent third person singular simply enough, as above, expressing the generalisation 'everything else', is not so easy.[41] With the LKB system, we could do this by making use of the type hierarchy, so that **pernum** has subtypes **3sg** and **non3sg**, where **3sg** has the constraint shown above, and all the other combinations are represented as subcases of **non3sg**. This is shown below (I have omitted the definition of the types **per** and **num** and their subtypes since these should be obvious).

```
pernum := *top* &
[ PER per,
  NUM num ].


non3sg := pernum.

1sg := non3sg &
[ PER 1st,
  NUM sg ].

1pl := non3sg &
[ PER 1st,
  NUM pl ].

2sg := non3sg &
[ PER 2nd,
  NUM sg ].

2pl := non3sg &
[ PER 2nd,
  NUM pl ].

3sg := pernum &
[ PER 3rd,
  NUM sg ].

3pl := non3sg &
[ PER 3rd,
  NUM pl ].
```

---

[41]One option is to add negation to the formalism, and indeed this has been done with some variants of constraint-based formalisms, but it turns out to have lots of formal and implementational consequences. Rather than go through the issues here, I will simply observe that, in general, I take the position that the formalism should only be expanded when there's a real gain in expressiveness and naturalness for a reasonable number of linguistic examples, and negation fails this test in my opinion.

This sort of encoding is used in Sag and Wasow (1999). However it still isn't very elegant in the LKB formalism because nothing blocks the following structure:

(5.60)
$$\begin{bmatrix} \textbf{non3sg} \\ \text{PER } \textbf{3rd} \\ \text{NUM } \textbf{sg} \end{bmatrix}$$

The reason for this is that the notion of well-formedness used in the LKB is actually quite weak and it does not preclude a TFS of a type **t** being well-formed even if it is not compatible with any of the subtypes of **t**.[42]

At this point, however, we have almost duplicated the feature information in the type hierarchy and we should ask what the justification is for the use of the features PER and NUM. The specific question is whether the grammar ever requires reentrancies between number information separately from person information. If there is no reason in the grammar to make these pieces of information independently accessible, we can simply drop the internal structure, and use a hierarchy of atomic types instead, as shown below.

```
pernum := *top*.
non3sg := pernum.
1sg := non3sg.
1pl := non3sg.
2sg := non3sg.
2pl := non3sg.
3sg := pernum.
3pl := non3sg.
```

We have thus removed the problem of the inconvenient TFS shown in 5.60, by the expedient of removing the internal structure. In practice, this style of agreement coding works well for English: a somewhat more complicated version of this hierarchy is used in the large-scale LinGO English Resource Grammar.

5.3.4 Type hierarchies versus disjunction

Although many feature structure based languages allow disjunctive feature structures, this is avoided in the LKB system. Arbitrary disjunction can result in a computationally intractable system. Furthermore, support for disjunction causes computational overheads in TFS representation and operations such as unification, even if the particular grammar contains no disjunctions. Disjunction within TFSs is never essential,

---

[42] Again, there are typed feature structure formalisms in which this could be ruled out, including some earlier versions of the LKB system, but there's a large cost in computational efficiency. See §5.6 for some further discussion.

given that the type system can be set up in a way which allows degrees of underspecification.[43] Note that the least common supertype of two types might be more general than their disjunction. In this case, a new type may have to be created to get the required level of specification. For instance, with the agreement example above, the type **non3sg** allowed us to encode the equivalent of a disjunction between **1sg**, **1pl**, **2sg**, **2pl** and **3pl**, which excluded **3sg**. Without **non3sg**, the join of **1sg** and **1pl** would be **pernum**.

If arbitrary disjunctions of types are required, then simulation of this effect in the type hierarchy is very cumbersome. To allow all possible disjuncts of the six types for agreement would require an additional 56 types. However, grammars do not generally seem to require this, so the effect of the non-equivalence of disjunction and join can be exploited to allow a more precise specification of a language. For example, Pollard and Sag (1994) give a description of the inflection of German adjectives such as *klein*, in which the following values for case are given as possibilities: **nom**, **acc**, **gen**, **dat**, **nom** ∨ **acc**, **gen** ∨ **dat**, *unspecified*. Encoding this in the type system, as shown below rather than making use of disjunction, directly expresses the restriction that only these values are available and that the following are not: **nom** ∨ **gen**, **acc** ∨ **dat**, **nom** ∨ **acc** ∨ **gen**, **nom** ∨ **acc** ∨ **dat**, **nom** ∨ **gen** ∨ **dat**, **acc** ∨ **gen** ∨ **dat**.

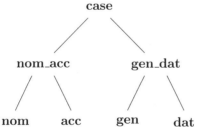

So, although it is sometimes annoying to have to explicitly add types, the payoff is that the specification of the language is more precise and generalizations are captured which are only implicit in a grammar with disjunctions.

5.4 Simple semantics and generation

The next grammar we will look at includes a simple specification of semantics. The idea of encoding semantics in TFSs is to build it up in

---

[43]This is not to say there is no disjunction in the system: lexical ambiguity, for instance, is a form of disjunction. However, this is a case where two TFSs are regarded as alternatives, rather than there being disjunctive information within a single structure.

parallel with the syntax, taking advantage of the richness of the TFS encoding. The result is generally taken to be equivalent to some more conventional semantic representation. Here we will look at a very simple language which is a first-order predicate logic without variables and quantifiers. The only connective we'll assume is conjunction ($\wedge$). All openclass words will contribute predicates. For instance, $[\text{dog}(c) \wedge \text{bark}(c)]$ is a valid logical form in this logic, where c is a constant.

Of course, to deal with natural language properly, we require variables, quantifiers, negation and so on. The simplified logic is purely for expository purposes: TFSs can be used to encode sentence semantics that utilizes quantifiers and higher order predicates and indeed most grammar engineers regard a richer language as essential.

A couple more points should be mentioned, before we go on to look at the grammar in detail. Although we won't deal with quantifiers, we will give *this*, *these*, etc a semantic representation: for instance *this dog* will correspond to $[\text{this}(c) \wedge \text{dog}(c)]$. This approach is adopted for convenience: it is clear that demonstratives don't have a simple truth-conditional meaning but I will not discuss their representation here. As is quite usual in computational semantics, I will use a representation for verbs that involves an argument that stands for the *event* or *situation*. For instance, the logical representation for *this dog barks* is equivalent to $[\text{this}(c) \wedge \text{dog}(c) \wedge \text{bark}(e, c)]$, where e is the event. The explicit representation of the event allows us to treat more phenomena within the very constrained semantic representation language. For instance, *this dog barks near the cat* is equivalent to $[\text{this}(c) \wedge \text{dog}(c) \wedge \text{bark}(e, c) \wedge \text{near}(e, c') \wedge \text{the}(c') \wedge \text{cat}(c')]$, which states that it is the barking event which is near the cat entity.[44]

I will take the sentence *this dog barks near the cat* to be an example of *modification*: the PP *near the cat* modifies the VP *barks* but the result is still a VP. We haven't seen a grammar that can deal with modification up to now, but I have added modification to grammar **g7sem** because it demonstrates some interesting facets of the semantics. In order to allow modification, I have added a new feature MOD, which like SPR and COMPS takes a list of **sign**s. However, unlike SPR and COMPS, MOD is an appropriate feature for **pos**, so it is in the substructure that HEAD leads to rather than at the top level of the **sign**. The **head-modifier-rule** combines a modifier with the modifiee, but the modifiee, rather than the modifier is the (syntactic) head daughter. The only **lexeme**s which have a non-empty MOD list in this grammar are **prep-lxm**s. Because MOD is

---

[44]Given the nature of barking, this will also presumably imply that the dog is near the cat, but this isn't true for all predicates.

a HEAD feature, this automatically means that the Pps also have non-empty MOD lists too, which allows them to combine with both Ns and VPs via the head-modifier-rule. I will not go through modification in detail, but leave it up to the reader to investigate it by trying out example sentences with PPs in the LKB.

5.4.1   Flat semantics in TFSs: Minimal Recursion Semantics

In order to represent semantics in the TFSs, we introduce a new toplevel feature SEM which has a value of type **semantics**. The semantic struc-ture for the sentence *this dog barks* is shown below:

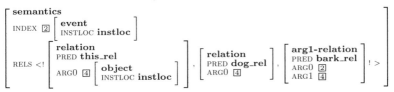

semantics has two appropriate features, INDEX and RELS. INDEX has a **sement** as a value: a **sement** may be an ordinary **object** variable or an **event** variable. **sement**s have the appropriate feature INSTLOC: this is needed for the generator, but can otherwise be ignored. INDEX itself plays no role in the semantics of a complete sentence — I will discuss its role in construction of the semantics in §5.4.2 and §5.4.3 below.

The feature RELS takes a difference list as a value. We build up the semantics as a list of *elementary predications*, by which we mean a combination of a single predicate with its arguments. Conjunction between elementary predications (EPs) is implicit. For instance, the TFS above shows a RELS list which is equivalent to [this(c) ∧ dog(c) ∧ bark(e, c)] although there isn't anything directly corresponding to the ∧s. We call this generic approach, where EPs are built up as a list, *flat semantics*, because there is no embedding of predications: the specific version that we're assuming is a simplified form of Minimal Recursion Semantics (MRS).[45] Each EP consists of a **relation**. A **relation** has the appropriate feature PRED which has a string value corresponding to the predicate symbol. It also has the appropriate feature ARG0, which is the event argument for verbs (e.g., the equivalent of the e in bark(e, c)) or the single argument for ordinary nouns (e.g., the c in dog(c)). Predicates which require more arguments introduce them with ARG1, ARG2 and ARG3, as required. Each subtype of **relation** has a fixed number of arguments. Equivalence of arguments is implemented by coindexation:

---

[45]It might seem that that flat semantics is only viable for very simple semantics: in fact it turns out that we can extend it to full-blown higher-order predicate calculus, as discussed in Copestake et al (1999).

for instance, the ARG1 of the relation corresponding to *bark* in the TFS above is coindexed with the ARG0 of *dog*.

The equivalence between the RELS list and the conventional representation should be fairly obvious. For each EP, the predicate corresponds to the value of PRED, the arguments correspond to the values of ARG0 ... ARG3, and the values themselves are represented by constant symbols, with coindexed values being represented by the same constant. The list of EPs is conjoined by ∧.

The LKB contains a semantic module which has code which can carry out this conversion. An option is available to show the conventional representation (although without explicit ∧) via the menu item **Indexed MRS** on the small parse tree display.

In order to manipulate semantic representations expressed in TFSs, for display or other purposes, the grammar engineer needs to specify a set of parameters which describe the way the semantics is encoded. The general idea of flat semantics is somewhat hardwired into the LKB's semantics module and generator, although not into the parser or underlying TFS manipulation code. But within this there is a lot of variation in the way that semantics can be encoded. The specific parameters needed for grammar g7sem are stored in mrsglobals.lsp which is loaded as part of the script. The first four lines of the script file are as follows:

```
(lkb-load-lisp (this-directory) "globals.lsp")
(lkb-load-lisp (this-directory) "mrsglobals.lsp")
(lkb-load-lisp (this-directory) "user-fns.lsp")
(lkb-load-lisp (this-directory) "user-prefs.lsp")
```

The parameters in mrsglobals.lsp are described in §9.2.6, but most users will probably find it easiest to work with the parameters from an exisiting grammar, at least initially.

5.4.2 Semantics in the lexicon: linking

Two sample lexical entries from g7sem are shown below:

```
dog := noun-lxm &
[ ORTH.LIST.FIRST "dog",
  SEM.RELS.LIST.FIRST.PRED "dog_rel" ].
```

```
chase := trans-verb &
[ ORTH.LIST.FIRST "chase",
  SEM.RELS.LIST.FIRST.PRED "chase_rel" ].
```

Now that we've added semantics, lexical entries are a triple consisting of orthography, semantic predicate symbol and lexical type (e.g., "dog", "dog_rel" and **noun-lxm**). The revised lexical type encodes both syntax and a skeleton semantic structure. Even with much more complex grammars, lexical entries will generally only consist of these three pieces of information.[46] Here I have made predicates be string-valued, which is the simplest option, because we do not have to declare them as explicit types. An alternative is for them to be normal types, which allows a hierarchy of predicates to be constructed, but I won't discuss that option here.

It is necessary to ensure that the correct semantic arguments are coindexed when phrases are combined. In linguistics, the connection between syntax and semantics in the lexicon is generally referred to as *linking*. In order to implement a very simple approach to linking in this grammar, the semantic argument positions are coindexed with the appropriate part of the syntax in the lexical entry. For instance, the sign for *chase* is shown below (INSTLOC and some other irrelevant details are omitted):

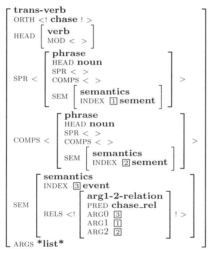

Note that the ARG1 value of the EP (i.e., SEM.RELS.ARG1) is reentrant with the INDEX of the specifier (SPR.SEM.INDEX) and that the ARG2 value is reentrant with the INDEX of the complement (COMPS.FIRST.SEM.INDEX). This illustrates a general principle: access to the semantics of a phrase is always via its INDEX slot — there is never a case where the RELS list is accessed directly. In fact the RELS list is simply there in order to build up the list of EPs. The term for building up semantics is *composition*: I will describe how this works in this grammar in the next section.

---

[46] At least, lexical entries for words: in a comprehensive grammar, lexical entries are also required for certain classes of phrase, and these require additional information to be supplied. But this is not considered in this book.

5.4.3 Composition

One of the reasons for adopting a flat semantic representation is that composition rules can be made very simple. Composition in **g7sem** always involves the following constraints:

1. The RELS of the mother of the phrase is the append of the RELS of the daughters.
2. One phrase has one or more syntactic slots (MOD, SPR or COMPS) filled by the other daughters. I'll refer to this phrase as the *semantic head*. The semantic head coindexes its argument positions with the INDEXs of the other daughters.
3. The INDEX on the phrase as a whole is coindexed with the INDEX of the semantic head daughter.

The semantic head daughter is not necessarily the same as the syntactic head, though in the case of this grammar, it is only in **head-modifier-phrase**s that the syntactic and semantic heads are different. For instance, in the phrase *dogs near the aardvark*, the N *dogs* is the syntactic head, but the PP *near the aardvark*, which is the modifier, is the semantic head.

The structure below shows a partial parse structure for the phrase *chase the cat*.

$$
\begin{bmatrix}
\textbf{binary-head-initial} \\
\text{ORTH} <! \textbf{ chase the cat } ! > \\
\text{HEAD } \boxed{1}\textbf{verb} \\
\text{SPR } \boxed{2} < \begin{bmatrix} \text{SEM.INDEX } \boxed{3} \end{bmatrix} > \\
\text{COMPS} < > \\
\text{SEM}\begin{bmatrix}
\text{INDEX } \boxed{4}\textbf{event} \\
\text{RELS} <!\begin{bmatrix}\text{PRED } \textbf{chase\_rel} \\ \text{ARG0 } \boxed{4} \\ \text{ARG1 } \boxed{3} \\ \text{ARG2 } \boxed{5}\end{bmatrix}, \begin{bmatrix}\text{PRED } \textbf{the\_rel} \\ \text{ARG0 } \boxed{5}\end{bmatrix}, \begin{bmatrix}\text{PRED } \textbf{cat\_rel} \\ \text{ARG0 } \boxed{5}\end{bmatrix}! >
\end{bmatrix} \\
\text{ARGS} < \begin{bmatrix}\textbf{plur-verb} \\ \text{HEAD } \boxed{1} \\ \text{SPR } \boxed{2} \\ \text{COMPS } < \boxed{6} > \\ \text{SEM.INDEX } \boxed{4}\end{bmatrix}, \boxed{6}\begin{bmatrix}\textbf{binary-head-second} \\ \text{SEM.INDEX } \boxed{5}\end{bmatrix} >
\end{bmatrix}
$$

You should try parsing some other sentences in the LKB system and look at the parse trees and parse charts so you see how the coindexation works.

5.4.4 The LKB generator

The LKB includes a generator which can produce strings on the basis of semantic representations. As we saw in Chapter 2, this can be demonstrated by parsing a sentence and then generating from its semantics. With grammar **g7sem**, the logical form $[\text{this}(c) \wedge \text{dog}(c) \wedge \text{chase}(e, c, c') \wedge \text{the}(c') \wedge \text{cat}(c')]$ generates the following strings:

```
this dog chased the cat
```

```
this dog chased the cats
this dog chases the cat
this dog chases the cats
```

You can try this for yourself: parse any one of the sentences above and then generate from it (via the **Generate** option on the menu accessible from clicking on the small parse tree): you should see a window containing those four sentences. You can click on the sentences to show their TFSs and corresponding tree. The reason we get *chases* as well as *chased*, and *cats* as well as *cat*, is that the semantic structures we're producing don't encode any information about tense or plurality. Obviously the grammar could be extended to include this, though I won't consider that here.

Generation in the LKB is formally as described in §4.1 except that we need to think more carefully about the notion of a start structure. In §4.1, I mentioned that the start sign for generation should be taken as including the semantics for the sentence, in contrast to parsing, where the start sign has the string for the sentence. We need to refine this idea, however, because some possible variations in the form that the logical representation takes are uninteresting semantically. For instance, the following two expressions are semantically equivalent, because $\wedge$ is commutative and associative:

1. $[\text{this}(c) \wedge \text{dog}(c) \wedge \text{chase}(e, c, c') \wedge \text{the}(c') \wedge \text{cat}(c')]$
2. $[\text{cat}(c') \wedge \text{chase}(e, c, c') \wedge \text{dog}(c) \wedge \text{the}(c') \wedge \text{this}(c)]$

When we translate an expression into TFSs, as we would have to do in order to construct a start structure, these issues become important, because although the order is not semantically significant, only some orderings will correspond to those allowed by the grammar. For grammar **g7sem**, the first order is the only one that is allowed, because EP order follows the order of the corresponding words in the sentence. This is clearly problematic, since when we're generating we don't know the word order. The solution adopted to this in the LKB system is formally equivalent to automatically specifying a set of alternative start structures which cover all possible permutations of the EPs. This is very easy to define, because we're using flat semantics. The actual technique obviously has to be a bit more efficient, as we'll discuss below.

There are a lot of issues about controlling generation and deciding on whether logical forms should be regarded as equivalent for generation purposes. For instance, logically *canis familiaris* might be equivalent to *dog* since they denotes the same set of entities (if we assume the most natural sense of *dog*). But we probably don't want to make that substitution in a generator. Unfortunately discussing these issues here would take us too far afield, so we'll just assume that order of EPS is the minimum variation we have to allow for, and leave it as an open issue whether any more complicated equivalences should also be covered.

The generation algorithm used in the LKB is *chart generation*, which is a variant of chart parsing. You will see a menu option to display the generator chart under **Generate** on the LKB Top window (note that this requires that the Expanded LKB Top window is visible — see **Expand menu** under **Options**). Generation algorithms are less well explored than parsing algorithms and chart generation is still somewhat experimental, so you will not find it discussed in standard textbooks. The LKB's algorithm is described in detail in Carroll et al (1999). In outline, chart generation consist of the following steps:

1. A set of possible lexical signs is constructed on the basis of the input logical form. For instance, if the logical form contains [this(c)∧ dog(c) ∧ bark(e, c)] then in **g7sem** we'll want the set of signs to include **this**, **bark** and **dog**. Any cases of argument equivalence are represented by putting a freshly-generated unique *constant type* into the relevant argument position on both signs: the function of the feature INSTLOC is to act as a location for that constant. Constant types are like string types in that they do not have to be declared explicitly and are mutually incompatible, but constant types are always automatically generated by the LKB system. They always have names that begin with a %, and those constructed for generation are all prefixed by **%instloc**. Figure 14 shows the **lexemes** for [this(c)∧dog(c)]. Note that the constant type **%instloc2** is the same on both the TFSs.

2. Morphological and lexical rules which are compatible with the semantics are applied, extending the set of signs. In this grammar, since we do not have a representation for tense and plurality, the extended set of signs will include those corresponding to *dogs*, *dog*, *bark*, *barks* and *barked*.

3. The set of signs is used to generate by parsing. We don't know the order in which the signs should appear in the eventual string of course, so the chart can't be indexed by string position as in chart parsing. Instead it is indexed by the value of the INDEX feature.[47] Each phrase that is built up is checked to ensure that it could contribute to the input semantics before being added to the chart.

4. The final TFS's semantics is checked to ensure it is compatible with the input semantics, as well as undergoing the usual check for the

---

[47]It is really a happy accident that this feature is called INDEX, since the name was first used well before chart generation was developed.

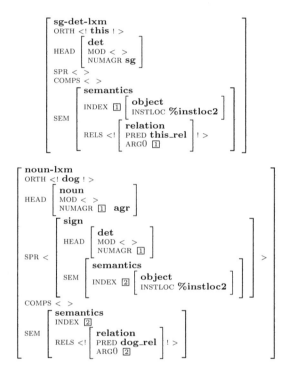

FIGURE 14 Lexical signs constructed for generation

syntactic start structure. The semantic compatibility check here and in the previous step is independent of the actual order of the EPs in RELS, which achieves the same effect as having a set of start structures that cover all permutations, but is reasonably efficient.

Because the lexical and morphological rules are applied in a slightly different way in the generator compared to the parser, the trees shown for generated sentences do not have nodes for the lexical and morphological rules. In other respects, parse trees and generator trees are identical.

In order to use a grammar for generation, the lexicon has to be indexed by its semantics (by default, the LKB lexicons are just indexed by orthography). The script file for g7sem has a command in it to do the indexing:

(index-for-generator)

The equivalent command can also be run from the extended LKB top menu: the command **Index** is under **Generate**.

5.4.5 Using the LKB as an interface to other systems

The LKB includes some code which can be used to link flat semantic representations to other systems, such as theorem provers. For instance, given a suitable semantic representation, a parsed sentence in the LKB can be converted to a normal form representation that can be added to a knowledge base or used to query it. Similarly, a logical form from another system can be converted into a flat semantic representation which can be passed to the LKB generator. Detailed discussion of this is, unfortunately, outside the scope of the present book, but further details will be made available on the LKB website.

5.4.6 *Reusability: optional exercise

The suggested exercise is to write a categorial grammar in the style of catgram that includes lexical rules, morphological rules, modifiers and semantics. You should reuse as much code from grammar g7sem as possible. In particular, you should attempt to build the same semantic structures and to use the same names for the lexical types. If you do this thoroughly, you should be able to use the lexicon file from g7sem for your grammar. You could also try splitting the types file in g7sem so that you have one file which is shared between g7sem and your grammar.

This is an open-ended exercise, so no sample solution is given. You should expect it to be quite time-consuming, but attempting it will help make sure that you understand the LKB and the TFS formalism. It should also demonstrate that TFS grammars can be written in such a way that they are substantially reusable. Investigating alternative approaches to syntax does not require that the entire grammar be thrown away and rebuilt. In particular, a large amount of effort in grammar engineering concerns the construction of a lexicon, but a well-designed lexicon can be shared between different grammars. The task is not as straightforward as it may appear from the rather simple grammars we've been looking at: multiword expressions, such as idioms and verb particle constructions are problematic, for instance. However, the general principle holds and trying this exercise should help you learn to write reusable grammars.

Another form of grammar sharing arises when encoding a grammar for another language. Obviously in this case the lexicon will be different, but it is often possible to share a substantial proportion of the type system. This is illustrated by the Spanish grammar developed by Ana Paula Quirino Simoes and distributed from the same website as the LKB system. The Spanish grammar shares a considerable proportion of its architecture with the textbook grammar for English (i.e., the grammar based on Sag and Wasow (1999)). Writing a small grammar for another language is also a very good way of learning about grammar engineering.

5.5   Long distance dependencies

The final grammar we will look at, `g8gap`, extends `g7sem` by including a treatment of topicalization. For instance it will parse:

> That cat, this dog chased.

I have put a comma in the sentence above, to try and make the topicalization clearer, but the LKB's default tokenizer actually ignores all punctuation. It is clear that sentences like this are only acceptable in marked contexts and with particular intonation, but for current purposes I will follow the usual tradition in formal linguistics and assume topicalized sentences are grammatical without worrying about these aspects.[48]

The reason for looking at topicalization here is that it is the simplest phenomenon to implement that involves long-distance dependencies: one of the most interesting things about feature structure grammars is their ability to handle long distance dependencies without any extra formal

---

[48]The example below may help if you are sceptical that such sentences are ever acceptable:

> My dog didn't chase your cat. Kim's cat, my dog did chase, I admit, but not yours.

However is a bit difficult to come up with natural sounding topicalized sentences when restricting oneself to cats, dogs and aardvarks, so if you haven't been previously convinced by a linguistics course or a textbook, please suspend disbelief for now!

machinery. Topicalization can involve multiple embedded clauses, for instance:

That cat, I believe you know your dog chased.

Long-distance dependencies are also needed to handle questions. For instance:

Which cat did you think this dog chased near the aardvark?

Questions, however, raise a range of other issues, so to keep things relatively simple, g8gap only deals with topicalization.

In traditional linguistics, the phrase *that aardvark* in 5.61 would be said to have been 'moved' from its normal position after *chased* and before *near*.

(5.61)   That aardvark, this dog chased near the cat.

Even in formalisms like the LKB's, which don't have a concept of movement, the term *gap* is often used to refer to the position of the 'missing' phrase — the gap is indicated by the underlined space in 5.62:

(5.62)   That aardvark, this dog chased _ near the cat.

The term *trace* is also used.

In what follows, I will give a very brief description of the treatment of long distance dependencies in grammar g8gap. The intention is that the reader works out some of the details by experimenting with the system. Further details of linguistic issues are given by Sag and Wasow (1999). Gaps are handled in g8gap by using a feature GAP which acts as a location for the specification of the 'missing' element. The gap is generally 'passed up' (i.e., loosely speaking, if a daughter in a phrase contains a gap, the phrase as a whole will share the gap) but can be discharged by a topicalized phrase. The tree in 5.63 illustrates this for 5.61.

(5.63)

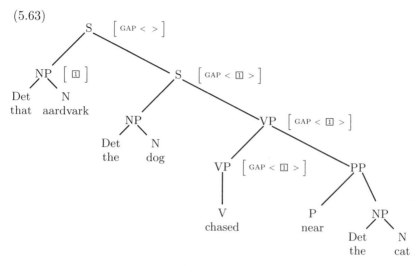

GAP is empty on most phrases (e.g., on all nodes where it is not explicitly shown in 5.63) but is instantiated by rules which are variants of the head-complement-rule but which take one less non-head-daughter than usual, corresponding to the gap position. The rule that applies to *chased* in this example and instantiates the GAP is the head-gap-rule-1: this is the rule represented by the VP node above *chased* in 5.63. The value of the single element of GAP is coindexed with the slot for the missing argument: i.e., the single member of *chased*'s COMPS for this example. The description for head-gap-rule-1 is shown below:

```
head-gap-rule-1 := unary-head-initial-startgap &
[ GAP <! #1 !>,
  SPR #spr,
  ARGS < word &
          [ SPR #spr,
            COMPS < #1 > ] > ].
```

There are two other gap introduction rules in g8gap which are for the ditransitive verbs.

Phrases with non-empty gap lists behave like normal phrases in nearly every respect. The two exceptions are:

1. A phrase with a non-empty GAP can be the head daughter of the head-filler-rule. The other daughter is the *gap filler*: it is unified with the single element of the GAP list. In the tree in 5.63, the phrase corresponding to this rule is represented by the uppermost S node which has an empty gap. The gap in the lower

S node, S [GAP < ☐ >], is filled by the NP [☐].

2. The start structure in **g8gap** specifies that GAP is a difference list with a LIST value which is the empty list. This means that *the dog chased near the aardvark* is not accepted as a complete sentence.[49]

In **g8gap**, GAP is implemented as a difference list. The grammar doesn't actually allow multiple gaps, but it is convenient to use the difference list encoding because it allows gaps to be passed up from any one of the daughters of a phrase. For instance, the type **binary-rule-passgap** constrains all the binary rules that don't either introduce or fill a gap so that the value of GAP on the phrase is the difference-list append of the gaps of the daughters.

```
binary-rule-passgap := binary-rule &
[ GAP [LIST #gfront, LAST #gtail ],
  ARGS < [ GAP [LIST #gfront, LAST #gmiddle ] ],
         [ GAP [LIST #gmiddle, LAST #gtail ] ] > ].
```

In the tree diagrams used in linguistic textbooks, phrases with gaps are often indicated by a slash followed by the abbreviation for the missing phrase. This is shown in the tree in 5.64.

(5.64)

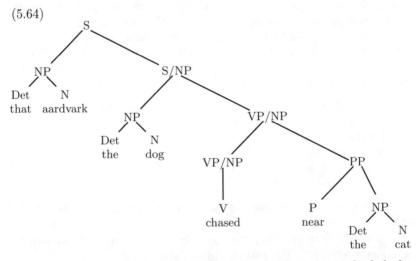

In **g8gap**, the tree nodes are labelled with hyphens instead of slashes (e.g., VP-NP): the more complicated tree node labelling scheme discussed in §8.14 can produce the conventional notation with less verbosity

---

[49]You might want to think about the difference between specifying GAP.LIST as the empty list, and specifying GAP to be the type ***null-dlist*** which specifies that its LIST is equal to its LAST.

than the g8gap/parse-nodes.tdl version.

Variants of the gap passing approach have been implemented in many computational feature structure grammars. I leave it to the reader to investigate g8gap and see how this mechanism works in detail. Notice that the file test.items contains some topicalized sentences. If you try these, you will discover that g8gap overgenerates in some respects: you might like to work out how you could prevent this. If you try generating, you will discover that both topicalized and non-topicalized variants are produced, because the semantics is identical (modulo the ordering of the elementary predications).

5.6 *A final note on formalism issues

This section is intended specifically for readers who have worked through Sag and Wasow (1999) or Pollard and Sag (1994) and who are worried by the discrepancies between that and the LKB formalism. All other readers should skip this! For convenience, the following discussion refers to Sag and Wasow (1999) alone, though much of it also applies to the earlier book.

At the beginning of Chapter 3, I noted that Sag and Wasow (1999) use the term 'typed feature structure description' where I use typed feature structure. The nearest counterpart in the LKB formalism to a Sag and Wasow 'typed feature structure' is a structure which is maximally specific with respect to types: that is, the type on each of its nodes is a leaf type in the hierarchy (this isn't necessarily a maximally specific structure in our terms because of the possibility of coindexation). But for Sag and Wasow, a 'typed feature structure description' and a 'typed feature structure' are two distinct classes of entity.

Sag and Wasow are (fairly informally) following a formal tradition in which the objects used in grammars are thought of as descriptions corresponding to sets of structures. King (1989, 1994) provides a logic which is probably the closest to Sag and Wasow's assumptions, but there are some discrepancies and Sag and Wasow themselves don't specify a logic. So, in this section, I will try and informally indicate what the issues are.[50] I will refer to Sag and Wasow's 'typed feature structure descriptions' as SWDs and their 'typed feature structures' as SWSs. To recapitulate the most salient representational issues from Sag and Wasow (1999):

1. A maximally specific valid SWD corresponds to a singleton set containing one SWS.

---

[50]Carpenter (1992:Ch4) provides a much more detailed discussion of how the two alternative formal approaches may be compared.

2. The unification of two SWDs corresponds to intersection of the set of SWSs they each describe.

3. A SWD that corresponds to the empty set is invalid.

As we have seen, the LKB's formalization does not involve the description/feature structure duality (except in the very limited sense that TDL descriptions have to correspond to TFSs), and I have described it in terms of operations on partial orders rather than on sets of underlying objects. Since SWSs are never actually written by a grammar writer or produced by an implementation, this is not in itself as important a distinction as it may seem. One of the things that does hang on it, however, is negation: it turns out it doesn't make sense to negate a TFS within the style of formalism the LKB uses (although see Carpenter (1992) on inequalities), but it is formally possible to negate a description in some feature structure logics (though there are efficiency consequences when this is implemented)[51]. In contrast, the LKB formalism can be extended to allow defaults (e.g., Lascarides and Copestake, 1999), but it is not clear that this is compatible with SWDs.[52]

A related issue is the notion of constraint. In §3.5, we expressed constraints on types in terms of TFSs and subsumption. In §4.1, we utilized another form of constraint on the well-formedness of phrases, also in terms of TFSs and subsumption, but expressed specifically with respect to a set of grammar rules and lexical entries. In the sort of approach Sag and Wasow discuss, these two concepts can be combined, assuming that the type hierarchy is set up correctly. Any node in a SWD is constrained not only by any locally specified or inherited constraint for its type, but also by all the constraints on that type's subtypes and so on, down to the maximally specific types. This follows from the definition of validity of SWDs, since they must always correspond to one or more SWSs. For instance, if **phrase** has subtypes, then a SWD of **phrase** is only valid if it can be specialized to a SWD of one of those subtypes, because a SWS could not contain a node of type **phrase** (since it is not maximally specific). If we assume that the SWDs associated with subtypes of **phrase** correspond to the notion of a grammar rule and that the SWDs associated with subtypes of **lexeme** are lexical entries, then parsing and generation both follow as part of the process of checking that a description is valid. If we start off with a root phrase with specified orthography, the **lexemes** where the constraint resolution bottoms out

---

[51]Sag and Wasow do not use negation, however.

[52]Sag and Wasow actually assume defaults, but as far as is known at the time of writing, these would have to be formalized as an abbreviatory notation for describing SWDs, i.e., a description of a description, at least if SWDs were defined in terms of King's (1994) formalism.

will have to correspond to that orthography (assuming, of course, that we constrain the orthography of phrases to be some combination of the orthography of the daughters).

This is, in some ways, very elegant because it means that every object of interest in the description of a grammar is a type constraint and all notions of well-formedness follow from the same fundamental principle.[53] This notion of type constraint is strong enough that useful operations such as append can be defined as types. Thus we can combine lists without using difference lists.

Unfortunately, from the grammar engineering perspective, the idea of a completely uniform approach to constraints is unattractive. It means that checking a grammar rule for well-formedness involves generating a sentence! As grammar engineers, we need a simple notion of constraint that quickly tells us whether we have inadvertently put a feature in the wrong place or specified a value that clashes with a parent. For practical purposes, we could perhaps separate the simple well-formedness of SWDs from the recursive notion that we need for parsing and generation. But even so, debugging grammars using this style of constraint is very difficult. With conventional parsing, debugging a large grammar can be hard, but once a problematic phrase has been established, the reason for unexpected success or failure is determined by a small number of TFSs that are always accessible via the parse chart. The search space for full recursive constraint resolution is far larger and more diffuse, so it is very difficult to determine where a problem has occurred, even with a small grammar. For similar reasons, processing such grammars is not efficient. There is a widespread belief that HPSGs are computationally intractable, which is largely due to the assumption that they have to be processed by general constraint resolution. A few systems have been developed that can parse and generate in this manner (in fact one early version of the LKB implemented such an approach as an alternative to the conventional parser), but no such system has yet achieved practical efficiency on realistic grammars.

---

[53]Sag and Wasow do not actually say that lexical entries are subtypes of **lexeme**, but that they are *instances* of **lexeme**. What they mean by 'instance' is not entirely clear, but lexical entries clearly have to be something other than simple SWDs of type **lexeme** because it is necessary to ensure that only structures which combine the right bundle of orthography, syntax and semantics are valid **lexeme**s: we must not allow a SWS which has the orthography "dog" paired with the semantics **cat_rel**, for instance. Formally, we could either have types **dog-lexeme**, **cat-lexeme** and so on, or we could define the maximally specific subtypes of **lexeme** as the disjunction of all the lexical entries of that type. The former move is a little counterintuitive but the latter step would make lexical entries different classes of object from the rest of the grammar, which Sag and Wasow say they avoid.

So, at least at the moment, the only practical option for an implementation is to treat different classes of constraint differently. Once this move is made, the implications of the gap between the formalisms narrows considerably. As shown in §4.3.1, the formal definition of a grammar in the LKB could be expressed in terms of type constraints, by specifying extra constraints on **phrase**, rather than making the formalism require the richer notion of constraint uniformly. But it turns out that nearly everything in Sag and Wasow (1999) can be implemented even with the more restricted notion of parsing that is actually used in the current LKB system and defined in §4.1. The only substantial problem is the binding theory, which requires a messy series of unary rules (lexical or grammar rules) to simulate the constraints proposed.

In fact, in HPSG generally, very little use is made of the power of general constraint resolution.[54] The notation used encourages this situation: it is still common to use trees and to describe phrasal constraints with the arrow notation in HPSG, for instance. The use of AVMs, instead of combinations of logical formulae, also restrict what is expressed. Furthermore some of the underlying assumptions of the HPSG framework, in particular the *locality principle*, explicitly constrain the way phrasal constraints behave. So far, most researchers have found that the difficult problems with implementing HPSG analyses in the LKB are centered around word order issues (especially for languages with a freer word order than English). But these require relatively straightforward extensions to the notion of parsing (see §4.3.1) rather than full constraint resolution.

To summarize, there are two approaches to developing a formalization of typed feature structure that is to be combined with a usable implementation. The approach that the LKB takes is to have a relatively restricted base language and add to it in order to be able to define the notion of a grammar. In particular, recursive constraints are limited to phrases. An approach which takes Sag and Wasow (1999) or Pollard and Sag (1994) more literally starts off with a very powerful language and then restricts it in order to have something that can be used for processing. It is clear that implementations based on the two approaches could meet in the middle and be equivalent, in which case the choice of formalism to describe them is not particularly important to the gram-

---

[54]Here, as throughout this book, I am using *power* informally. Since the LKB, along with most unification-based systems, can be used to encode grammars which are Turing equivalent, in principle any analysis could be simulated by a sufficiently complex LKB grammar. But such a simulation could be unrecognizable and unusable: the point here is that, in my experience, a practically useful restatement that avoids general constraint resolution can usually be found.

mar engineer. Despite the fact that most work in HPSG talks about the full-blown constraint formalism, if formalism is discussed at all, the linguistic analyses proposed generally don't exploit this power. However, the best position for the middle ground has not yet been established. For now, the grammar engineer needs to know the formalism that underlies the system that's being used, but it is a matter of taste whether to regard it as an expedient adopted for computational purposes or as a principled attempt at capturing linguistic intuitions!

5.7 Summary

The aim of this chapter was to describe how we can exploit typed feature structures to build more interesting grammars. The particular approach discussed was a simplified version of HPSG, loosely based on Sag and Wasow (1999). In the course of this, a lot of material was introduced which illustrated more ways in which the typed feature structure formalism could be exploited, including techniques which can be used for the description of long distance dependencies. The only new pieces of machinery discussed were lexical and morphological rules. A very brief account was provided of flat semantics and generation. The exercises were designed to help you learn more about grammar building and to start using the LKB system to write your own grammars.

5.8 Further information

This section contains suggestions for further reading, pointers to other systems and some information about the history of unification-based approaches in general and the LKB system in particular.

5.8.1 Further reading

There is a great deal more to be said about variations in formalisms and systems which cannot be discussed in an introductory book. There is also a large literature on parsing and generation algorithms in various frameworks, which I haven't even begun to discuss. The references here are definitely not intended as a complete bibliography, but simply pointers into the literature which the interested reader can use as stepping stones to other work. I have also listed some papers which are specifically connected with the LKB system. Most LKB related papers are available from the LKB website.

For a general introduction to parsing, including unification based systems, see Jurafsky and Martin (2000). This describes chart parsing and unification in detail and references a large number of publications on algorithms and efficiency. Flickinger et al (2000) contains a range of papers on recent experiments with efficiency in typed feature structure systems, all using the LinGO ERG as a common baseline.

There have been several formalizations of feature structures and unification: Kasper and Rounds (1986, 1990) being one of the best known and the closest to the approach assumed here. Carpenter (1992) provides the best introduction to formal work on typed feature structures.

Unfortunately, there is very little published work specifically about grammar implementation or grammar engineering. One notable exception is Butt et al (1999) which discusses many aspects of the wide-coverage grammars implemented for several languages in the ParGram parallel grammar development project. This uses the Lexical Functional Grammar (LFG) linguistic framework with the Xerox Linguistic Environment (XLE) but contains much that is relevant to other approaches. Alshawi (1992) discusses the Core Language Engine (CLE), including some information about the grammar. Pulman (1996) describes some techniques for exploiting unification within the CLE and similar systems — unfortunately nothing comparable has been published for typed feature structure systems, although Flickinger (2000) discusses some techniques used in the LinGO ERG. Some of the websites listed in §5.8.2 have links to detailed documentation of various grammars.

Linguistic frameworks which use feature structures are quite popular, especially LFG and HPSG, and many books and papers are published each year about analyses of different phenomena in a range of languages (many of which are published by CSLI Publications). Often these accounts can be exploited by the grammar engineer, although the process of incorporating them into a computational grammar is rarely completely straightforward. The LinGO project webpage (http://lingo.stanford.edu) lists several papers which describe analyses which have been implemented in the LKB system, mostly as part of the LinGO ERG.

5.8.2 Other available systems

I have listed here systems which are downloadable freely, at least for research purposes, and which have been used for substantial grammars or by multiple research groups. Links to the websites are given on the LKB webpage — I have not listed URLs here, since these may change. The main developer(s) (as listed on the websites) are shown — please see the websites for details of the systems and associated publications.

General purpose feature structure systems:

- ALE — The Attribute Logic Engine (Bob Carpenter and Gerald Penn). One of the most widely used typed feature structure systems, originally developed by Bob Carpenter, now taken over by

Gerald Penn.

- HDRUG (Gertjan van Noord). A Prolog grammar development environment with a graphical user interface and a range of grammars, parsers and generators.
- TFS (Martin Emele). One of the first typed feature structure systems.
- ConTroll — (Detmar Meurers, Thilo Götz, Gerald Penn and others). An implementation based on King (1994).
- LiLFeS (Takaki Makino and others). This system has been used with large scale grammars for Japanese and English and also in experiments with the LinGO ERG.
- ProFIT — Prolog with Features, Inheritance and Templates (Gregor Erbach). ProFIT is an extension of Prolog rather than a full grammar development environment, but has been used by a number of researchers for implementing grammars.
- PC-PATR (Stephen McConnel): a version of Shieber's original PATR system.

The following are not general purpose feature structure systems, but are listed here because they share a relatively similar philosophy and have been used with substantial grammars.

- Grok (Gann Bierner and Jason Baldridge). This provides a range of tools for implementing systems based on a version of categorial grammar.
- XTAG project (XTAG Research Group). Grammar, parser and other tools for the Lexicalized Tree Adjoining Grammar (TAG) framework.
- Grammar Writer's Workbench (John Maxwell and Ron Kaplan). This is the predecessor to the XLE.
- DATR (Roger Evans, Gerald Gazdar and Bill Keller). DATR is intended for development of lexicons, but it can be integrated with unification-based parsing systems.

The following systems require a license, or have other restrictions on availability:

- ALEP — Advanced Language Engineering Platform. Freely available to sites within the EU.
- ANLT — Alvey Natural Language Tools (Ted Briscoe, John Carroll, Claire Grover and others). Morphological analyser, parser and wide coverage English grammar and lexicon integrated in a grammar development environment that has been widely used for NLP research since the 1980s. Not free.

5.8.3 Some highlights of the history of unification-based approaches

To do justice to the history of typed feature structures would require a long essay. What I have tried to do in this short section is just to give references to some of the work which has been most influential from my viewpoint — it is inevitably a somewhat subjective list. Carpenter (1992) gives a more detailed introduction with many more references. Note, in particular, his discussion of the links between TFSs and representation languages used in artificial intelligence. See also Jurafsky and Martin (200:442–444).

Martin Kay is generally credited with inventing feature structure unification (Kay, 1979). Kay's use of unification was independent from the use of unification in logic programming, which had also been advocated for encoding natural language (Colmerauer, 1970). However, more recent work has often connected the two, and definite clause grammars (DCGs, Pereira and Warren, 1980) underlie several Prolog-based implementations of feature structure systems. PATR-II (Shieber et al, 1983) was probably the most widely used system based on untyped feature structures. Many of the basic ideas about grammar engineering in unification-based systems can be traced back to work on PATR-II, though much of this was unpublished or only circulated as technical reports. D-PATR provided a full grammar development environment based on the PATR-II formalism (Kartunnen, 1986). Shieber's (1986) account of unification-based grammars is the standard reference to the use of feature structure grammars in computational linguistics.

Several well-known linguistic frameworks have made use of feature structures and unification. Kay's own Functional Grammar (also known as Functional Unification Grammar, FUG) did not become widely popular with linguists but Lexical Functional Grammar (LFG: Bresnan and Kaplan, 1982), which also adopted the use of unification in the 1970s, has been very influential in formal and computational linguistics. Generalized Phrase Structure Grammar (GPSG, Gazdar et al, 1985) used a rather restricted feature formalism (with some complications, especially the use of defaults), although the GPSG use of principles and meta-rules is often mirrored by the use of type constraints in later work in HPSG. The account of gap passing discussed in §5.5 goes back to Gazdar (1981). Unification Categorial Grammar (UCG: Zeevat et al, 1987) is a version of categorial grammar that uses feature structures (the grammar given in catgram is actually descended from UCG via Sanfilippo (1993)). HPSG (Pollard and Sag: 1987,1994) is the linguistic framework which is most often associated with the use of TFSs (although note that types are referred to as *sorts* in much work on HPSG). More recently, a lot of work on lexical semantics, especially that based on the Generative Lexicon (Pustejovsky, 1995), has adopted TFSs.

As I described in §5.3.2, the most important use of types in almost all LKB grammars is to capture linguistic generalizations. Inheritance networks of various sorts have been popular for knowledge representation in artificial intelligence for a long time. The idea that hierarchical structuring should be a major organizational principle for lexicons had been advocated by several people by the mid-1980s, but the approach which most directly influenced typed feature structure grammars came out of research on NLP at Hewlett-Packard Laboratories and at Stanford University (Flickinger et al, 1995, Flickinger, 1987). This drew on concepts from object-oriented programming, but type constraints were later adopted by some researchers as a way of formalizing lexical inheritance. Although Flickinger made some use of default inheritance, which is not supported by standard type systems, type constraints are nevertheless able to capture the main insights of this work.

The first formal work on typed feature structures was Aït-Kaci (1984). Smolka (1989) extended this and showed the relationship between the different approaches to formalizing feature structures. King (1989, 1994) and Carpenter (1992) provided formalizations that were intended to capture the ideas that were informally described by Pollard and Sag (1987). The LKB's approach is heavily based on Carpenter's, although not identical to it.

The first typed feature structure implementation to be used for grammar development was probably that described by Calder (1987) and Moens et al (1989) for UCG. The TFS system (Emele and Zajac, 1990; Emele, 1994) was another early typed feature structure system which was notable because it adopted full constraint resolution for parsing and generation. Since then, many systems have been implemented. Typed feature structure formalisms certainly didn't supersede the untyped constraint-based systems: as mentioned above, the ANLT, the XLE and the CLE have all been used for large scale grammar development without types. In fact, it is only recently that typed feature structure based grammars of similar scale have been implemented.

5.8.4 Historical notes on the LKB system

The first version of the LKB system was implemented at Cambridge University in early 1991 as part of the ACQUILEX project. At this point, LKB stood for Lexical Knowledge Base — the LKB system was a tool for building LKBs. Although it incorporated a parser, this was of secondary importance, since it was mostly intended as a way of verifying lexical entries. The most extensive documentation of that version of the system is Copestake (1993): published papers include Copestake (1992) and several papers in Briscoe et al (1993). de Paiva (1993) gives a formalization of the LKB's notion of types and constraints.

The LKB system was intermittently developed for a number of years, mostly as a way of trying out various theoretical ideas. In 1998, much more intensive development effort began, primarily in order to make the system usable for large scale grammar development as part of the LinGO project at Stanford University. The new version of the system was implemented by John Carroll, Rob Malouf, Stephan Oepen and myself. Some of the main changes compared to the older system are:

1. Extensive efficiency improvements, so the system is capable of parsing reasonable length sentences with a large grammar (see Malouf et al (2000) and Oepen and Carroll (2000a,b)).

2. A new treatment of defaults, based on Lascarides and Copestake (1999), with an efficient algorithm developed by Malouf.

3. Automatic computation of greatest lower bounds in the type hierarchy. (The algorithm currently used for this is due to Callmeier.)

4. Integration with Oepen's [incr tsdb()][55] test suite machinery (Oepen and Flickinger, 1998), which is briefly described in §8.13.

5. Integration with MRS semantics (Copestake et al, 1999).

6. Tactical generation from MRS input (Carroll et al, 1999).

7. Many new user interface features.

Some of the earlier functionality was not incorporated into the new version: either because it was considered obsolete or simply because of shortage of time. The system is under continual development: users are encouraged to check the LKB web page for details of recent modifications.

---

[55]This is not a LaTeX macro error, it really is what the system is called!

Part II

LKB User Manual

6

LKB user interface

In this chapter, I will go through the details of the menu commands and other aspects of the graphical user interface. The default LKB top menu window has six main menus: Quit, Load, View, Parse, Debug and Options. If you select **Options / Expand menu**, you will obtain a menu which has nine main menus: Quit, Load, View, Parse, MRS, Generate, Debug, Advanced and Options. The expanded menu also makes more submenus available, and makes minor changes to one or two of the basic submenus. You can revert to the basic LKB top menu window with **Options / Shrink menu**.

The first section in this chapter describes the commands available from the LKB top menu, while subsequent sections describe the windows which display different classes of LKB data structure, which have their own associated commands. Specifically, the sections are as follows:

1. Top level commands. Describes the commands associated with the menus and submenus in the order in which they appear in the expanded LKB top menu window. Items which are only in the expanded menu are marked by *.
2. Type hierarchy display
3. Typed feature structure display
4. Parse output display
5. Parse tree display
6. Chart display

All the data structure windows have three buttons in common: **Close**, **Close all** and **Print**. **Close** will close the individual window, **Close all** will close all windows of that class — e.g., all type hierarchy windows etc. **Print** produces a dialog that allows a Postscript file to be output which can then be printed. Printing directly to a printer is not implemented yet.

Most commands that produce some output do so by displaying a new window. A few commands output short messages to the LKB interaction window. A small number of less frequently used commands send output to the standard Lisp output, which is generally the emacs *common-lisp* or *shell* buffer, if the LKB is being run from emacs, and the window from which the LKB was started, if emacs is not being used. These commands are all ones from which a large amount of text may be produced and the reason for outputting the text to an emacs buffer is that the results can be searched (it is also considerably faster than generating a new window).

Very occasionally it is useful to be able to run a command from the Lisp command line (i.e., the window where prompts such as LKB(1): appear, which will be the same window as the one displaying the standard Lisp output). This is easier to do using emacs, since commands can be edited.

Because the LKB system is under active development, some minor changes may be made to the commands described here and additional functionality will probably appear. Documentation for any major modifications will be available from the website.

6.1 Top level commands

The top level command window is displayed when the LKB is started up. In this section the LKB menu commands will be briefly described in the order in which they appear in the expanded interface. Commands which are only in the expanded menu are indicated by ∗. To switch between versions, use the **Shrink menu** or **Expand menu** commands under **Options**.

6.1.1 Quit

Prompts the user to check if they really do want to quit. If so, it shuts down the LKB.

6.1.2 Load

The commands allow the loading of a script file (see §4.5.1 and §8.2) and the reloading of the same file (normally this would be done after some editing). A script is used initially to load a set of files, and can be reloaded as necessary after editing.

Complete grammar This prompts for a script file to load, and then loads the grammar. Messages appear in the LKB interaction window.

Reload grammar This reloads the last loaded script file. Messages appear in the LKB interaction window.

Other reload commands∗ The other reload commands are for reloading parts of the grammar — they should not be used by the inexperienced, since under some conditions they will not give the correct behaviour. If something unexpected happens after using one of these commands, always reload the complete grammar.

6.1.3 View

These commands all concern the display of various entities in the grammar. Many of these commands prompt for the name of a type or entry. If there are a relatively small number of possibilities, these will be displayed in a menu.

Type hierarchy Displays a type hierarchy window.

Prompts for the highest node to be displayed. If the type hierarchy under this node is very large, the system will double-check that you want to continue (generally, large hierarchies won't be very readable, so it's probably not worth the wait). The check box allows 'invisible' types, such as glbtypes, to be displayed if set. Details of the type hierarchy window are in §6.2.

Type definition Shows the definition of a type constraint plus the type's parents.

Prompts for the name of a type. If the type hierarchy window is displayed, scrolls the type hierarchy window so that the chosen type is centered and highlighted. Displays the type's parents and the constraint specification in a TFS window: details of TFS windows are in §6.3.

Expanded type Shows the fully expanded constraint of a type.

Prompts for the name of a type. If the type hierarchy window is displayed, scrolls the type hierarchy window so that the chosen type is centered and highlighted. Displays the type's parents and the full constraint on the type.

Lex entry The expanded TFS associated with a lexical entry (or parse node label or start structure etc). The command is used for entries other than lexical entries to avoid having a very long menu.

Prompts for the identifier of a lexical entry (or a parse node label or start structure). Displays the associated TFS.

Word entries All the expanded TFSs associated with a particular orthographic form.

Prompts for a word stem. Displays the TFSs corresponding to lexical entries which have this stem.

Grammar rule Displays a grammar rule.

Prompts for the name of a grammar rule (if there are sufficiently few rules, they are displayed in a menu from which the name can be chosen), displays it in a TFS window.

Lexical rule Displays a lexical or morphological rule.

Prompts for the name of a lexical rule (if there are sufficiently few rules, they are displayed in a menu from which the name can be chosen), displays its TFS in a window.

All words∗ Displays a list of all the words defined in the lexicon, in the emacs `*common-lisp*` buffer (if emacs is being used), otherwise in the window from which the LKB was launched.

6.1.4 Parse

Parse input This command prompts the user for a sentence (or any string), and calls the parser (tokenizing the input according to the user-defined function `preprocess-sentence-string`, see §9.3). A valid parse is defined as a structure which spans the entire input and which will unify with the TFS(s) identified by the value of the parameter `*start-symbol*`, if specified (i.e., the start structure(s), see §4.5.6 and §4.2). (Note that `*start-symbol*` may be set interactively.) If there is a valid parse, a single window with the parse tree(s) is displayed (see §6.4).

It is sometimes more useful to run the parser from the Lisp command line interface, since this means that any results generated by post-processing will appear in an editor buffer and can be searched, edited and so on. It may also be useful to do this if you have to use emacs to enter diacritics. The command `do-parse-tty` is therefore available — it takes a string as an argument. For example:

```
(do-parse-tty "Kim sleeps")
```

The normal graphical parse output is produced.

Redisplay parse Shows the tree(s) from the last parse again.

Show parse chart Shows the parse chart for the last parse (see §6.6).

Batch parse This prompts for the name of a file which contains sentences on which you wish to check the operation of the parser, one sentence per line (see the file `test.items` in the sample grammars). It then prompts for the name of a new file to which the results will be output. The output tells you the number of parses found (if any) for each sentence in the input file and the number of passive edges, and gives a time for the whole set at the end. This is a very simple form of test suite: vastly more functionality is available from the [incr tsdb()] machinery which can be run in conjunction with the LKB (see §8.13).

Compare∗ This is a tool for treebanking: it displays the results of the last parse, together with a dialog that allows selection / rejection of rule applications which differ between the parses. It thus allows comparison of parses according to the rules applied. It is intended for collection of data on preferences but can also be useful for distinguishing between a large set of parse results. Specifying that a particular phrase is in/out will cause the relevant parse trees to be indicated as possible/impossible and the other phrases to be marked in/out, to the extent that this can be determined. The parameter *discriminant-path* can be set to identify a useful discriminating position in a structure: the default value corresponds to the location of the key relation in the semantic structure used by the LinGO ERG.

The treebanking tool is under active development at the time of writing, and so a full description is not given here. Documentation will be made available via the LKB webpage.

6.1.5 MRS*

The MRS commands relate to semantic representation, but they assume a particular style of semantic encoding, as is used in the LinGO ERG. The grammars discussed in this book use a simplified version of MRS. MRS is briefly discussed in §5.4 and §8.11. MRS output can be displayed in various ways by clicking on the result of a parse in the compact parse tree representation (see §6.4) or displayed in the main editor window (*common-lisp* buffer or Listener), as controlled by the **Output level** command below. The parameterisation for MRS is controlled by various MRS-specific files, discussed in §8.11.

Load munger The term *munger* refers to a set of rules which manipulate the MRS in application-specific ways. Loading a new set of rules will overwrite the previously loaded set. Most users should ignore this.

Clear munger Deletes the munger rules.

Output level Allows the user to control the MRS output which is sent to the standard Lisp output (an emacs buffer, if emacs is being used). This command is provided since with large structures it is often more convenient to look at MRS output in emacs rather than in the MRS windows displayed by clicking on a tree in the parse output window. The default output level is NONE, but this may be changed by the grammar-specific MRS globals files.

- NONE
- BASE: a bracketed representation of the tree, plus an underspecified MRS, generally quite similar to the TFS representation.
- SCOPED: the scoped forms corresponding to the underspecified structure produced by the grammar. If no scoped forms can be produced, warning messages are output. If there are a large num-

ber of scoped forms, only a limited number are shown, by default. Because scoping can be computationally expensive, there is a limit on the search space for scopes: this is controlled by `mrs::*scoping-call-limit*`.

6.1.6 Generate∗

The generator was described in §5.4 and a few more details are given in §8.12, but is currently in a fairly early stage of development. It operates in a very similar manner to the parser but relies on the use of flat semantics such as MRS, thus it will only work with grammars that produce such semantics. Before the generator can be used, the command **Index** must be run from this menu. Alternatively, the script can include the command:

`(index-for-generator)`

At the moment, there is no interactive way of entering an MRS input for the generator other than by parsing a sentence which produces that MRS and then choosing **Generate** from the appropriate parse window.

Redisplay realisation Redisplays the results from the last sentence generated.

Show gen chart Displays a window showing a chart from the generation process (see §6.6). Note that the ordering of items on the chart is controlled by their semantic indices.

Load heuristics Prompts for a file which should contain a set of heuristics for determining null semantics lexical items (see §8.12).

Clear heuristics Clears a set of heuristics, loaded as above.

Index Indexes the lexicon and the rules for the generator. This has to be run before anything can be generated. Any error messages are displayed in the LKB top window.

6.1.7 Debug

Check lexicon Expands all entries in the lexicon, notifying the user of any entries which fail to expand (via error messages in the LKB top window). This will take a few minutes for a large lexicon. An alternative for small grammars is to have the command

`(batch-check-lexicon)`

in the script file.

Find features' type∗ Used to find the maximal type (if any) for a list of features (see §3.5.8 for a discussion of maximal types). Prompts for a list of features. Displays the maximum type in the LKB interaction window. Warns if feature is not known.

Print chart / Print parser chart∗ Displays the chart (crudely) to the standard Lisp output (e.g., the emacs buffer). This can be useful as an alternative display to the parse chart window, especially with very large charts.

Print generator chart∗ As above, but for the generator.

6.1.8 Advanced∗

Tidy up This command clears expanded lexical entries which are stored in memory. If accessed again they will be read from file and expanded again.

Expansion of a large number of word senses will tend to fill up memory with a large number of TFSs. Most commands which are likely to do this to excess, such as the batch parser, actually clear the TFSs themselves, but if a lot of sentences have been parsed interactively and memory is becoming restricted this option may be useful.

Create quick check file The check path mechanism constructs a filter which improves efficiency by processing a set of example sentences. It is discussed in more detail in §8.3.1.

The command prompts for a file of test sentences and an output file to which the resulting paths should be written. This file should subsequently be read in by the script. Note that constructing the check paths is fairly time-consuming, but it is not necessary to use a very large set of sentences. The mechanism is mildly grammar-specific in that it assumes the style of encoding where the daughters of a rule are given by an ARGS list — see §8.3.1 for details.

6.1.9 Options

Expand/Shrink menu Changes the LKB top menu so that the advanced commands are added/removed.

Set options Allows interactive setting of some system parameters. Note that the values of the boolean parameters are specified in the standard way for Common Lisp: that is, t indicates true and nil indicates false. I will not go through the parameters here: Chapter 9 gives full details of all parameters, including those that cannot be altered interactively.

If a parameter file has been read in by the script (using the load function load-lkb-preferences) the parameter settings are saved in the same file. Otherwise the user is prompted for the name of a file to save any preference changes to. This file would then have to be specified in the script if the changes are to be reloaded in a subsequent session.

Usually the preferences file is loaded by the script so that any preferences which are set in one session will be automatically saved for a subsequent session with that grammar. (In the cases of 'families' of grammars, the user-prefs file may be shared by all the grammars in the family.) The user should not need to look at this file and should not edit it, since any changes may be overwritten.

Save display settings Save shrunkenness of TFSs (see the description of **Shrink/Expand** in §6.3).

Load display options Load pre-saved display setting file.

6.2 Type hierarchy display

By default, a type hierarchy is displayed automatically after a grammar is loaded (though this default must be turned off for grammars that use very large numbers of types, see §9.1.1). The type hierarchy can also be accessed via the top level command **Type hierarchy** in the **View** menu, as discussed above in §6.1.3.

The top of the hierarchy, that is the most general type, is displayed at the left of the window. The window is scrollable by the user and is automatically scrolled by various **View** options. Nodes in the window are active; clicking on a type node will give a menu with the following options:

Shrink/Expand Shrinking a type node results in the type hierarchy being redisplayed without the part of the hierarchy which appears under that type being shown. The shrunk type is indicated by an outline box. Any subtypes of a shrunk type which are also subtypes of an unshrunk type will still be displayed. Selecting this option on a shrunk type reverses the process.

Type definition Display the definition for the constraint on that type (see §6.1.3, above).

Expanded type Display the expanded constraint for that type (see §6.1.3, above).

New hierarchy Displays the type hierarchy under the clicked-on node in a new window, via the same dialog as the top-level menu command. This is useful for complex hierarchies.

6.3 Typed feature structure display

Most of the view options display TFSs in a window. The usual orthographic conventions for drawing TFSs are followed; types are lowercased bold, features are uppercased. The order in which features are displayed in the TFS window is determined according to their order when introduced in the type specification file. For example, assume we have the following fragment of a type file:

```
sign := feat-struc &
 [ SYN *top*,
   SEM *top* ].
```

```
word := sign &
 [ ORTH string ].
```

then when a TFS of type **sign** is displayed, the features will be displayed in the order SYN, SEM; when a **word** is displayed the order will be SYN, SEM, ORTH. This ordering can be changed or further specified by means of the parameter *feature-ordering*, which consists of a list of features in the desired order (see §9.1.2).

The bar at the bottom of the TFS display window shows the path to the node the cursor is currently at.

Typed feature structure windows are active - currently the following operations are supported:

1. Clicking on the window identifier (i.e., the first item in the window) will display a menu of options which apply to the whole window.

   Output TeX Outputs the FS as LaTeX macros to a file selected by the user. The LaTeX macros are defined in avmmacros in the data directory.

   Apply lex rule Only available if the identifier points to something that might be a lexical entry. It prompts for a lexical or morphological rule and applies the rule to the entry. The result is displayed if application succeeds.

   Apply all lex rules This option is only available if the identifier points to something that might be a lexical entry. This tries to apply all the defined lexical and morphological rules to the entry, and to any results of the application and so on. (To prevent infinite recursion on inappropriately specified rules the number of applications is limited.) The results are displayed in summary form, for instance:

   dog + SG-NOUN_IRULE

   dog + PL-NOUN_IRULE

   Clicking on one of these summaries will display the resulting TFS.

   Show source Shows the source code for this structure if the system is being used with emacs with the LKB extensions. This is not available with all structures: it is not available for any entries which have been read in from a cached file.

2. Clicking on a reentrancy marker gives the following sub-menu:

   Find value Shows the value of this node, if it is not displayed at this point, scrolling as necessary.

   Find next Shows the next place in the display where there is a pointer to the node, scrolling as necessary.

3. Clicking on a type (either a parent, or a type in the TFS itself) will give a sub-menu with the following options:

   Hierarchy Scroll the type hierarchy window so that the type is centered. If the type hierarchy window is not visible, it will be redisplayed.

   Shrink/Expand Shrinking means that the TFS will be redisplayed without the TFS which follows the type being shown. The existence of further undisplayed structure is indicated by a box round the type. Atomic TFSs may not be shrunk. Shrinking persists, so that if the window is closed, and subsequently a new window opened onto that TFS, the shrunken status will be retained. Furthermore, if the shrunken structure is a type constraint, any TFSs which inherit from this constraint will also be displayed with equivalent parts hidden. For instance, if the constraint on a type has parts shrunk, any lexical entry which involves that type will also be displayed with parts hidden.

   If this option is chosen on an already shrunken TFS then the TFS will be expanded. Again this can affect the display of other structures.

   The shrunkenness state may be saved via and loaded via the **Save/Load display settings** commands on the **Options** menu (see §6.1.9).

   Show source Shows the source code for this structure if running from emacs with the LKB connection (not available with all structures).

   Type definition Display the definition for that type.

   Expanded type Display the expanded definition for that type.

   Select Selects the TFS rooted at the clicked node in order to test unification.

   Unify Attempts to unify the previously selected TFS with the selected node. Success or (detailed) failure messages are shown in the LKB Top window. See §6.3.1 for further details.

   Clicking on a type which is in fact a string, and thus has no definition etc, will result in the warning beep, and no display.

6.3.1 Unification checks

The unification check mechanism operates on TFSs that are displayed in windows. You can temporarily select any TFS or part of a TFS by clicking on the relevant node in a displayed window and choosing **Select** from the menu. Then to check whether this structure unifies with another, and to get detailed messages if unification fails, find the node corresponding to the second structure, click on that, and choose **Unify**. If the unification fails, failure messages will be shown in the top level LKB window. If it succeeds, a new TFS window will be displayed. This can in turn be used to check further unifications.

A detailed description of how to use this mechanism is in §7.4.1.

6.4 Parse output display

The parse output display is intended to give an easily readable overview of the results of a parse, even if there are several analyses. The display shows a parse tree for each separate parse, using a very small font to get as many trees as possible on the screen. Besides the standard **Close** and **Close all** buttons, the parse output display window has a button for **Show chart**: this has the same effect as the top-level menu command, it is just repeated here for convenience.

Clicking on a tree gives several options:

Show enlarged tree produces a full size parse tree window, as described in §6.5, with clickable nodes.

Highlight chart nodes will highlight the nodes on the parse chart corresponding to this tree. If the parse chart is not currently displayed, this option will bring up a new window (see §6.6 for details of the chart display).

Generate Tries to generate from the MRS for this parse. Note that in order to run the generator, the **Generate / Index** command must have been run. If generation succeeds, the strings generated are shown in a new window — clicking on the strings gives two options:

> **Edge** displays the tree associated with that realization,
>
> **Feature structure** displays the TFS associated with that realization.

If generation fails, the message 'No strings generated' will appear in the LKB interaction window.

MRS Displays an MRS in the feature structure style representation.

Prolog MRS Displays an MRS in a Prolog compatible notation (designed for full MRSs, rather than simplified MRSs).

Indexed MRS Displays an MRS using the alternative linear notation.

Scoped MRS Displays all the scopes that can be constructed from the MRS: warning messages will be output if the MRS does not scope.

6.5 Parse tree display

Parse trees are convenient abbreviations for TFSs representing phrases and their daughters. When a sentence is successfully parsed, the trees which display valid parses are automatically shown, but parse trees may also be displayed by clicking on any edge in a parse chart (see §6.6). The nodes in the parse tree are labelled with the name of the (first) parse node label which has a TFS which matches the TFS associated with the node, if such a label is present. The matching criteria are detailed in §4.5.7 and §8.14.

The input words are indicated in bold below the terminal parse tree nodes — if any morphological rules have been applied, these are indicated by nodes beneath the words if the parameter *show-morphology* is t, but not shown otherwise. Similarly, there is a parameter *show-lex-rules* which controls whether or not the lexical rule applications are displayed. Both these parameters may be set interactively, via the **Options / Set options** menu command.

Clicking on a node in the parse tree will give the following options:

Feature structure - Edge X (where X is the edge number in the parse chart) displays the TFS associated with a node. Note that if the parameter *deleted-daughter-features* is set, the tree will still display the full structure (it is reconstructed after parsing). See §8.3.3.

Show edge in chart Highlights the node in the chart corresponding to the edge. The chart will be redisplayed if necessary. Currently not available for a tree produced by the generator.

Rule X (where X is the name of the rule used to form the node) displays the TFS associated with the rule.

Generate from edge This attempts to generate a string from the MRS associated with this node. Behaves as the **Generate** command from the parse output display. Can give strange results if the node is not the uppermost one in the tree. Currently not available with a tree produced by the generator. Note that in order to run the generator, the **Generate / Index** command must have been run.

Lex ids This isn't a selectable option - it's just here as a way of listing the identifiers of the lexical entries under the node.

6.6   Chart display

The chart is a record of the structures that the LKB system has built in the course of attempting to find a valid parse or parses (see §4.2). A structure built by the parser and put on the chart is called an *edge*: edges are identified by an integer (*edge number*). By default, all edges that are displayed on the chart represent complete rule applications.

The chart window shows the words of the sentence to the left, with lines indicating how the structures corresponding to these words are combined to form phrases. Each node in the chart display corresponds to an edge in the chart. A node label shows the following information:

1. The nodes of the input that this edge covers (where the first node is notionally to the left of the first word and is numbered 0, just to show we're doing real computer science here).
2. The edge number (in square brackets).
3. The name of the rule used to construct the edge (or the type of the lexical item).

For instance, in the chart for the sentence *the dogs chased the cats*, the nodes for the input are numbered

$._0$ *the* $._1$ *dogs* $._2$ *chased* $._3$ *the* $._4$ *cats* $._5$

In the chart display resulting from parsing this sentence in the g8gap grammar, one edge is specified as:

2-5 [19] HEAD-COMPLEMENT-RULE-1

Thus this edge is edge number 19, it covers *chased the cats*, and was formed by applying the head-complement-rule-1.

The chart display is sensitive to the parameters *show-morphology* and *show-lex-rules* in a similar way to the tree display.

Moving the cursor over an edge in the chart displays the yield of the edge at the bottom of the window. Clicking on a word node (i.e., one of the nodes at the leftmost side of the chart which just show orthography) will select it. When at least one word is selected, all the edges that cover all the selected words are highlighted. Clicking on a word node again deselects it.

Clicking on an edge node results in the following menu:

Highlight nodes Highlights all the nodes in the chart for which the chosen node is an ancestor or a descendant. This option also selects the node so that it can be compared with another node (see **Compare**, below).

Feature structure Shows the TFS for the edge. Unlike the parse tree display, this represents the TFS which is actually used by the parser, see the discussion in §4.2. It is not reconstructed if `*deleted-daughter-features*` is used (see §8.3.3).

Rule X Shows the TFS for the rule that was used to create this edge

New chart Displays a new chart which only contains nodes for which the chosen node is an ancestor or a descendant (i.e., those that would be highlighted). This is useful for isolating structures when the chart contains hundreds of edges.

Tree Shows the tree headed by the phrase corresponding to this edge

Compare This option is only available if another node has been previously selected (using **Highlight Nodes**). The two nodes are compared using the parse tree comparison tool described in §6.1.4.

Unify This is only shown if a TFS is currently **Select**ed for the unification test — see §6.3.1.

7

Error messages and debugging techniques

This chapter is intended to help with debugging. There are two sorts of problems which arise when writing grammars in the LKB. In the first class, the system doesn't accept your grammars files and generates error messages. This type of problem is very irritating when you are learning how to use the system, but with experience, such problems generally become easy to fix. In this chapter, the error messages are explained in detail with references back to the chapters discussing the LKB formalism. The second type of problems are more difficult: the system doesn't give explicit error messages, but doesn't do what you want it to. Some debugging tools that can be used in this situation are described in §7.4.

7.1   Error messages

The formal conditions on the type hierarchy and the syntax of the language were detailed in Chapters 3 and 4. Here we will go through those conditions informally, and discuss what happens when you try and load a file in which they are violated. If you do not understand the terminology, please refer back to the earlier chapters.

Many examples of errors are given below: these all assume that we have made the minimal change to the g8gap grammar to make it match the structures shown. The errors are not supposed to be particularly realistic!

IMPORTANT NOTE: Look at all the messages in the LKB Top window when you load a grammar and always look at the first error message first! Error messages may scroll off the screen, so you may need to scroll up in order to do this. Sometimes errors propagate, causing other errors, so it's a good idea to reload the grammar after you have fixed the first error, rather than try and fix several at once, at least until you have gained familiarity with the system.

7.1.1 Type loading errors: Syntactic well-formedness

If the syntax of the constraint specifications in the type file is not correct, according to the definition in §4.4.6, then error messages will be generated. The system tries to make a partial recovery from syntactic errors, either by skipping to the end of a definition or inserting the character it expected, and then continuing to read the file. This recovery does not always work: sometimes the inserted character is not the intended one and sometimes an error recovery affects a subsequent definition. Thus you may get multiple error messages from a single error. The system will not try to do any further well-formedness checking on files with any syntactic errors. In the examples below, an incorrect definition is shown followed by the error message that is generated. All the definitions are based on g8gap/types.tdl.

Example 1: missing character

```
agr : *top*.
```

```
Syntax error at position 132:
Syntax error following type name AGR
Ignoring (part of) entry for AGR
Error: Syntax error(s) in type file
```

The error is caused by the missing = following the :. The error message indicates the position of the error (using emacs you can use the command goto-char to go to this position in the file). The number given will not always indicate the exact position of the problem, since the LKB's TDL description reader may not be able to detect the problem immediately, but is likely to be quite close. The system then says what it is doing to try and recover from the error (in this case, ignore the rest of the entry) and finally stops processing with the error message Syntax error(s) in type file (I will omit this in the rest of the examples).

Example 2: missing character

```
semantics := *top* &
[ INDEX index,
  RELS *dlist* .
```

```
Syntax error: ] expected and not found in SEMANTICS
at position 403
Inserting ]
```

In this example, the system tries to recover by inserting the character it thinks is missing, correctly here.

Example 3: missing character

```
semantics := *top* &
[ INDEX index
  RELS *dlist*] .
```

```
Syntax error: ] expected and not found in SEMANTICS
at position 389
Inserting ]
Syntax error: . expected and not found in SEMANTICS
at position 389
Inserting .
Syntax error at position 394
Incorrect syntax following type name RELS
Ignoring (part of) entry for RELS
```

Here the system diagnosed the error incorrectly, since in fact a comma was missing rather than a ']'. The system's recovery attempt doesn't work, and the error propagates. This illustrates why you should reload the grammar after fixing the first error unless you are reasonably sure the error messages are independent.

Example 4: coreference tag misspelled

```
unary-rule := phrase &
[ ORTH #orth,
  SEM #cont,
  ARGS < [ ORTH #orth, SEM #comt ] > ] .
```

```
Syntax error at position 821: Coreference COMT
only used once
```

In this example, the system warns that the coreference was only used once: it is assumed that this would only be due to an error on the part of the user.

Other syntax errors You may also get syntax errors such as the following:

```
Unexpected eof when reading X
```

eof stands for end of file — this sort of message is usually caused by a missing character.

7.1.2 Conditions on the type hierarchy

After a syntactically valid type file (or series of type files) is read in, the hierarchy of types is constructed and checked to ensure it meets the conditions specified in §3.2.

All types must be defined If a type is specified to have a parent which is not defined anywhere in the loaded files, an error message such as the following is generated:

```
AGR specified to have non-existent parent *TOPTYPE*
```

Although it is conventional to define parent types before their daughters in the file, this is not required, and order of type definition in general has no significance for the system. Note however that it is possible to redefine types, and if this is done, the actual definition will be the last one the system reads. If two definitions for types of the same name occur, a warning message will be generated, for instance:

```
    Type AGR redefined
```

Connectedness / unique top type There must be a single hierarchy containing all the types. Thus it is an error for a type to be defined without any parents, for example:

```
sign :=
[ ORTH *dlist*,
  HEAD pos,
  SPR *list*,
  COMPS *list*,
  SEM semantics,
  GAP *dlist*,
  ARGS *list* ].
```

Omitting the parent(s) of a type will cause an error message such as the following:

```
Error: Two top types *TOP* and SIGN have been defined
```

To fix this, define a parent for the type which is not intended to be the top type (i.e., **sign** in this example).

If a type is defined with a single parent which was specified to be its descendant, the connectedness check will give error messages such as the following for every descendant of the type:

```
NOUN not connected to top
```

(This situation is also invalid because cycles are not allowed in the hierarchy, but because of the way cycles are checked for, the error will be found by the connectedness check rather than the cyclicity check).

No cycles It is an error for a descendant of a type to be one of that type's ancestors. This causes an error message to be generated such as:

```
Cycle involving TERNARY-HEAD-INITIAL
```

The actual type specified in the messages may not be the one that needs to be changed, because the system cannot determine which link in the cycle is incorrect.

Redundant links This is the situation where a type is specified to be both an immediate and a non-immediate descendant of another type. For instance, suppose g8gap/types.tdl is changed so that **phrase** is specified as a parent of **unary-head-initial** even though it is already a parent of **head-initial** (and also a parent of **unary-rule** which is a parent of **unary-rule-passgap**):

```
unary-head-initial := unary-rule-passgap & head-initial
                    & phrase.
```

Then the error messages are:

```
Redundancy involving UNARY-HEAD-INITIAL
UNARY-HEAD-INITIAL: PHRASE is redundant
- it is an ancestor of UNARY-RULE-PASSGAP
UNARY-HEAD-INITIAL: PHRASE is redundant
- it is an ancestor of HEAD-INITIAL
```

The assumption is that this would only happen because of a user error. The condition is checked for because it could cause problems with the greatest lower bound code (see §4.5.3). After finding this error, the system checks for any other redundancies and reports them all.

7.1.3 Constraints

Once the type hierarchy is successfully computed, the constraint descriptions associated with types are checked, and inheritance and typing are performed to give expanded constraints on types (see §3.5).

Valid constraint description The check for syntactic well-formedness of the constraint description is performed as the files are loaded, but errors such as missing types which prevent a valid TFS being constructed are detected when the constraint description is expanded. For example, suppose the following was in the g8gap type file, but the type **foobar** was not defined.

```
sign := *top* &
[ ORTH *dlist*,
  HEAD foobar,
  SPR *list*,
  COMPS *list*,
  SEM semantics,
  GAP *dlist*,
  ARGS *list* ].
```

The first error messages are as follows:

```
Invalid type FOOBAR
Unifications specified are invalid or do not unify
Type SIGN has an invalid constraint specification
Type PHRASE's constraint specification clashes with
its parents'
```

Note that the error propagates because the descendants' constraints cannot be constructed either.

Another similar error is to declare two nodes to be reentrant which have incompatible values. For example:

```
sign := *top* &
[ ORTH *dlist*,
  HEAD pos & #1,
  SPR *list*,
  COMPS *list* & #1,
  SEM semantics,
  GAP *dlist*,
  ARGS *list* ].
```

The error messages would be very similar to the case above:

```
Unifications specified are invalid or do not unify
Type SIGN has an invalid constraint specification
Type PHRASE's constraint specification clashes with
its parents'
```

No Cycles TFSs are required to be acyclic in the LKB system (see §3.3): if a cycle is constructed during unification, then unification fails. In the case of construction of constraints, this sort of failure is indicated explicitly. For example, suppose the following is a type definition

```
wrong := binary-rule &
  [ ARGS < #1 & [ GAP < #1 > ], *top* > ] .
```

The following error is generated:

```
Cyclic check found cycle at < GAP : FIRST >
Unification failed - cyclic result
Unification failed: unifier found cycle at
                    < ARGS : FIRST >
Type WRONG has an invalid constraint specification
```

Consistent inheritance Constraints are constructed by monotonic inheritance from the parents' constraints. If the parental constraints do not unify with the constraint specification, or, in the case of multiple parents, if the parents' constraints are not mutually compatible, then the following error message is generated:

```
Type X's constraint specification clashes with its parents'
```

Maximal introduction of features As described in §3.5, there is a condition on the type system that any feature must be introduced at a single point in the hierarchy. That is, if a feature, F, is mentioned at the top level of a constraint on a type, t, and not on any of the constraints of ancestors of t, then all types where F is used in the constraint must be descendants of t. For example, the following would be an error because NUMAGR is a top level feature on both the constraint for **agr-cat** and **pos** but not on the constraints of any of their ancestors:

```
pos := *top* & [ MOD *list* ].

nominal := pos & [ NUMAGR agr ].

pseudonom := pos & [ NUMAGR agr ].
```

The error message is as follows:

```
Feature NUMAGR is introduced at multiple types
(POS PSEUDONOM)
```

To fix this, it is necessary to introduce another type on which to locate the feature. For example:

```
pos := *top* & [ MOD *list* ].

numintro := pos & [ NUMAGR agr ].

nominal := numintro.

pseudonom := numintro.
```

No infinite structures It is an error for a constraint on a type to mention that type inside the constraint. For example, the following is invalid.

```
*ne-list* := *list* &
 [ FIRST *top*,
   REST *ne-list* ].
```

The reason for this is that expansion of the constraint description would create an infinite structure (as discussed in §3.5.8). The following error message is produced:

```
Error in *NE-LIST*:
```

Type *NE-LIST* occurs in constraint for type
NE-LIST at (REST)

Similarly it is an error to mention a daughter of a type in its constraint. It is also an error to make two types mutually recursive:

```
foo := *top* &
[ F bar ].
```

```
bar := *top* &
[ G foo ].
```

The error message in this case is:

```
BAR is used in expanding its own constraint
                    expansion sequence: (FOO BAR)
```

Note that it *is* possible to define recursive constraints on types as long as they specify an ancestor of their type. For example, a correct definition of **list** is:

```
*list* := *top*.
```

```
*ne-list* := *list* &
 [ FIRST *top*,
   REST *list* ].
```

```
*null* := *list*.
```

Type inference — features There are two cases where typing may fail due to the feature introduction condition. The first is illustrated by the following example:

```
noun-lxm := lexeme &
[ HEAD [ NUMAGR #agr,
         INDEX *top* ],
  SPR < [HEAD det & [NUMAGR #agr],
         SEM.INDEX #index ] >,
  COMPS < >,
  SEM [ INDEX object & #index ] ].
```

Here, the feature INDEX is only defined for structures of type **semantics**. This type clashes with **pos** which is the value of HEAD specified higher in the hierarchy. This example generates the following error messages:

```
Error in NOUN-LXM:
  No possible type for features (NUMAGR INDEX) at
  path (HEAD)
```

A different error message is generated when a type is specified at a node which is incompatible with the node's features. For instance:

```
test := lex-item &
  [ ARGS < pos &
        [ HEAD *top* ] > ].
```

Error in TEST:
  Type of fs POS at path (ARGS FIRST) is incompatible
  with features (HEAD) which have maximal type SIGN

Type inference — type constraints The final class of error is caused when type inference causes a type to be determined for a node which then clashes with an existing specification on a path from that node. For instance:

```
nominal := pos & [ NUMAGR agr,
                   MOD <> ].
```

```
test1 := lexeme &
       [ ARGS < [ HEAD [ NUMAGR sg,
                         MOD < *top* > ]] > ].
```

Here the feature NUMARG in **test1** means that the node at the end of the path ARGS.FIRST.HEAD has to be of type **nominal**, but **nominal** specifies that its value for MOD is the empty list. The error message is

Unification with constraint of NOMINAL failed at
path (ARGS FIRST HEAD)

The **Debug / Find features' type** command on the expanded menu (§6.1.7) can be useful when trying to fix such problems.

7.2 Lexical entries

When lexicon files are loaded, they are generally only checked for syntactic correctness (as defined in §4.4.3) — entries are only fully expanded when they are needed during parsing or generation or because of a user request to view an entry. Thus when loading the lexicon, you may get syntactic errors similar to those discussed in §7.1.1 above, but not content errors since the TFSs are not expanded at load time. With a small lexicon, you can put the command batch-check-lexicon in the script file, in order to check correctness at load time, as was done with the example grammars for this book.

Incorrect lexical entries may therefore only be detected when you view a lexical entry or try to parse with it. The error messages that are obtained are very similar to some of those discussed for the type loading above, specifically:

Valid constraint description
No Cycles
Consistent inheritance
All types must be defined
Type inference — features
Type inference — type constraints

There is a menu option to do a complete check on a loaded lexicon for correctness: **Check lexicon** under **Debug**. See §6.1.7.

7.3 Grammar rules

Unlike lexical entries, grammar and lexical rules are always expanded at load time. Therefore you may get error messages similar to those listed above for lexical entries when the rules are loaded. Rules must expand out into TFSs which have identifiable paths for the mother and daughters of the rule (§4.4.4). For the grammars we have been looking at, the mother path is the empty path and the daughter paths are defined in terms of the ARGS feature. For example:

$$
\begin{bmatrix}
\textbf{binary-rule} \\
\text{ARGS} \begin{bmatrix} \text{FIRST } \textbf{sign} \\ \text{REST} \begin{bmatrix} \text{FIRST } \textbf{sign} \end{bmatrix} \end{bmatrix}
\end{bmatrix}
$$

Here the mother is given by the empty path, one daughter by the path ARGS.FIRST and another by the path ARGS.REST.FIRST. The mother path and a function which gives the daughters in the correct linear order must be specified as system parameters in the `globals` and `user-fns` files respectively (see §9.2.3 and §9.3).

7.4 Debugging techniques

If something unexpectedly fails to parse or if too many parses are produced, then isolating the problem can be quite tricky. The first thing to look at is usually the parse chart, though you may need to try and find a simpler sentence that produces the same effect, because the parse chart is difficult to navigate with long sentences. In extreme cases, the **Debug / Print chart** command may be useful (see §6.1.7). Note that the display at the bottom of the parse chart is extremely helpful in showing you what node you are looking at. Once the problem is isolated to a particular phrase, or small set of phrases, the unification checking mechanism discussed below becomes helpful.

Unfortunately debugging grammars is not a skill that can easily be described. After a while, a grammar writer becomes sufficiently familiar with a grammar that the likely sources of problems become reasonably obvious, but this takes practice. Most of the same principles apply as when debugging computer programs: in fact, the parse chart display can be thought of as a backtrace mechanism. Because the LKB is a development environment, we have kept the amount of grammar compilation to a minimum, so that the time taken to reload a grammar is fast, even with the LinGO ERG. For large scale grammars, the more sophisticated test suite techniques described in §8.13 are an invaluable tool. As soon as you start working with grammars, you should investigate the use of a version control system, such as the Concurrent Versions System (CVS): even if you are working by yourself, the ability to easily recover previous versions of grammars is extremely useful.

7.4.1 Use of the unification checking mechanism

As mentioned in §6.3.1 there is an interactive mechanism for checking unification via the TFSs that are displayed in windows. You can temporarily select any TFS or part of a TFS by clicking on the relevant node in a displayed window and choosing **Select** from the menu. Then to check whether this structure unifies with another, and to get detailed messages if unification fails, find the node corresponding to the second structure, click on that, and choose **Unify**. If the unification fails, failure messages will be shown in the top level LKB window. If it succeeds, a new TFS window will be displayed. This can in turn be used to check further unifications. Here we give a detailed example of how to use this.

Consider why *the dog chased* does not parse in the g8gap grammar. By looking at the parse chart for the sentence, you can determine that although the `head-specifier-rule` applies to give the phrase *the dog chased*, this isn't accepted as a parse. This must be because it fails the start symbol condition. To verify this:

1. Load the g8gap grammar.
2. Do **Parse / Parse input** on the dog chased. The parse will fail.
3. Do **Parse / Show parse chart**. A chart window will appear.
4. Click on the rightmost node in the chart, which is labelled HEAD-SPECIFIER-RULE and choose **Feature structure**. A TFS window will appear.
5. Click on the root node of the TFS (i.e., the node with type **binary-head-second-passgap**) and choose **Select**.
6. Display the TFS for `start` by choosing **Lex entry** from the **View** menu (`start` is not actually a lexical entry, but we wanted to avoid putting too many items on the **View** menu, so **Lex entry** is used as a default for viewing 'other' entries too).

7. Click on the root node of start (i.e., the node with the type **phrase**) and choose **Unify** from the pop-up menu.
8. Look at the LKB Top window. You should see:

```
Unification of *NE-LIST* and *NULL* failed at path
                              < GAP : LIST >
```

This demonstrates that the parse failed because the phrase contains a gap.

It is instructive to redefine start in g8gap as below, reload the grammar and retry this experiment:

```
start := phrase &
[ HEAD verb,
  SPR <>,
  COMPS <>,
  GAP <! !> ].
```

To check why a rule does not work, a more complex procedure is sometimes necessary, because of the need to see whether three (or more) TFSs can be unified. Suppose we want to check to see why the head-specifier rule does not apply to *the dog bark* in g8gap. This is relatively complex, since both the TFSs for the NP for *the dog* and the VP for *bark* will individually unify with the daughter slots of the head-specifier-rule. So we need to use the intermediate results from the unification test mechanism.

The following description details one way to do this.

1. Parse *the dog bark*. The parse will fail.
2. Select **Show chart** to show the edges produced during the failed parse.
3. Bring up the TFS window for the head-specifier-rule (uninstantiated), either via the **View Rule** command in the top window menu or by choosing **Rule HEAD-SPECIFIER-RULE** from a head-specifier-rule node in the chart.
4. Find the node in the head-specifier-rule window corresponding to the specifier (i.e., the node labelled **phrase** at the end of the path ARGS.FIRST). Click on it and choose **Select**.
5. Find the node for **the dog** in the parse chart window and choose **Unify**. (Note that this is a shortcut which is equivalent to displaying the TFS for the phrase via the **Feature Structure** option in the chart menu and then selecting **Unify** from the menu on the top node of the TFS displayed.) Unification should succeed and a new TFS window (titled Unification Result) will be displayed.

6. Find the node in the new Unification Result window corresponding to the head in the rule, i.e., the node at the end of the path ARGS.REST.FIRST, click on it and choose **Select**.

7. Click on the node in the parse chart for the VP for bark (i.e., the one labelled `head-complement-rule-0`) and choose **Unify**.

8. This time unification should fail, with the following message in the LKB Top window:

```
Unification of SG and PL failed at path
< SPR : FIRST : HEAD : NUMAGR >
```

Note that it is important to **Select** the node in the rule and **Unify** the daughters rather than vice versa because the result shown always corresponds to the initially selected TFS. We need this to be the whole rule so we can try unifying the other daughter into the partially instantiated result.

8

Advanced features

The previous chapters described the main features of the LKB system which are utilized by most of the distributed grammars. There are a range of other features which are in some sense advanced: for instance because they concern facilities for using the LKB with grammars in frameworks other than the variety of HPSG assumed here, or because they cover functionality which is less well tested (in particular defaults and generation), or because the features are primarily used for efficiency. These features are described in this chapter, which is a series of more or less polished notes about various aspects of the LKB. The intention is that the LKB website:

`http://cslipublications.stanford.edu/lkb.html`

will contain updated information. In any case, this chapter should only be read after working through the earlier material in the book in some detail. On the whole, it assumes rather more knowledge of NLP and of Lisp programming than I have in the earlier chapters.

8.1 Defining a new grammar

If possible, it is best to start from one of the sample grammars rather than to build a new grammar completely from scratch. However, when building a grammar in a framework other than HPSG, the existing sources may not be of much use. These notes are primarily intended for someone trying to build a grammar almost from scratch.

The first step is to try and decide whether the LKB system is going to be adequate for your needs. The system is not designed for building full NLP applications, though it can form part of such a system for teaching or research purposes, and has some utility for development of commercial applications. Grammars can be written and debugged in the LKB and then deployed using more efficient platforms such as Callmeier's PET system. There are some limitations imposed by the typed feature structure formalism.[56] There's no way of describing any form of transformation or movement directly, though, as we saw in §5.5, feature structure formalisms have alternative ways of achieving the same effects.

Even with respect to other typed feature structure formalisms, the LKB has some self-imposed limitations. As I have discussed, there is no way of writing a disjunctive or negated TFS (see §5.3.4). In many cases, grammars can be reformulated to eliminate disjunction in favour of the use of types which express generalisations. Occasionally it may be better to have multiple lexical entries or multiple grammar rules. The LKB does not support negation, although we do intend to release a version which incorporates inequalities (Carpenter, 1992) at some point. More fundamentally for HPSG, the system does not support set operations. Alternative formalisations are possible for most of the standard uses of sets. For example, operations which are described as set union in Pollard and Sag (1994) can be reformulated as a list append. We think we have good reasons for adopting these limitations, and that the LinGO ERG shows that a large scale HPSG grammar can be built without these devices, so the LKB system is unlikely to change in these respects.

There are other limitations which are less fundamental, though they might require considerable reimplementation. In these cases, it may be worth considering using the LKB system if you have some Lisp programming experience, or can persuade someone else to do some programming for you! For instance, the current system for encoding affixation is quite restricted. However, the interface to the rest of the system is well-defined, so it would be relatively easy to replace. As mentioned in previous chapters, the implementation of the parser has limitations for languages with relatively free word order. It would be possible to replace the parsing module to experiment with different algorithms. If you want to attempt any modifications like this, please feel free to email the LKB mailing list for advice (see the website for details of how to subscribe).

If you've decided you want to use the LKB system, then you should start by defining a very simple grammar. It will make life simpler if you copy the definitions for basic types like lists and difference lists from the existing grammars, so you do not have to redefine the global parameters unnecessarily. Similarly, if you are happy to use a list feature ARGS to

---

[56]The system can be used to build grammars which are Turing equivalent, so these comments aren't about formal power. There is a useful contrast between whether a language *supports* a technique, which means it supplies the right primitives etc, or merely *enables* its use, which means that one can implement the technique if one is sufficiently devious. What's of interest here is the techniques the LKB supports or doesn't support.

indicate the order of daughters in a grammar rule, you will not have to change the parameters which specify daughters or the ordering function. There are some basic architectural decisions which have to be taken early on. For instance, if you decide on an alternative to the **lexeme**, **word**, **phrase** distinction described in Chapter 5, this will affect how you write lexical and grammar rules.

As described in §4.5, you need to have distinct files for each class of object. You will have to write a script file to load your grammar files — the full details of how to do this are given below, but the easiest technique is to copy an existing script and to change the file names, then to look at the documentation if you find the behaviour surprising. To start off with, aim at a grammar which is comparable in scope to the `g5lex` grammar: i.e., use fully inflected forms, one or two grammar rules, a very small number of lexical entries and just enough types to make the structures well-formed (the `g5lex` grammar actually has more types than are strictly speaking needed, because it was designed to be relatively straightforward to extend using a predefined feature structure architecture).

There are many practical aspects to grammar engineering, most of which are similar to good practise in other forms of programming (see Copestake and Flickinger (2000) for some discussion). One aspect which is to some extent peculiar to grammar writing is the use of test suites (see e.g., Oepen and Flickinger, 1998). The LKB system is compatible with the [incr tsdb()] system, as discussed in §8.13. You should use a test suite of sentences as soon as you have got anything to parse, to help you tell quickly when something breaks. You should also adopt a convention with respect to format of description files right from the start: e.g., with respect to upper/lower case distinctions, indentation etc. Needless to say, comments and documentation are very important. As mentioned in §7.4, using a version control system such as CVS really does help when developing even moderately complex grammars. You may find the LKB's typing regime annoying at first, but we have found that it catches a large number of bugs quickly which can otherwise go undetected or be very hard to track down.

At some point, if you develop a moderate size grammar, or have a slow machine, you will probably start to worry about processing efficiency. Although this is partly a matter of the implementation of the LKB system, it is very heavily dependent on the grammar. For instance, the grammar based on Sag and Wasow (1999) (i.e., the 'textbook' grammar) is not a good model for anyone concerned with efficient processing, because it makes use of large numbers of non-branching rules. The LKB system code has been optimized with respect to the LinGO ERG, so it may well have considerable inefficiencies for other styles of grammar. For instance, the ERG has about 40 rules — grammars with hundreds of rules would probably benefit from an improvement in the rule lookup mechanism. On the other hand, we have put a lot of effort in making processing efficient with type hierarchies like the ERG's which contains thousands of types, with a reasonably high degree of multiple inheritance, and the LKB performs much better in this respect than many other typed feature structure systems.

8.2 Script files

Here is an example of a complex script file, as used for the CSLI LinGO ERG:

```
(lkb-load-lisp (parent-directory) "Version.lisp" t)
(lkb-load-lisp (this-directory) "globals.lsp")
(lkb-load-lisp (this-directory) "user-fns.lsp")
(load-lkb-preferences (this-directory) "user-prefs.lsp")
(lkb-load-lisp (this-directory) "checkpaths.lsp" t)
(lkb-load-lisp (this-directory) "comlex.lsp" t)
(load-irregular-spellings
    (lkb-pathname (parent-directory) "irregs.tab"))
(read-tdl-type-files-aux
    (list
        (lkb-pathname (parent-directory) "fundamentals.tdl")
        (lkb-pathname (parent-directory) "lextypes.tdl")
        (lkb-pathname (parent-directory) "syntax.tdl")
        (lkb-pathname (parent-directory) "lexrules.tdl")
        (lkb-pathname (parent-directory) "auxverbs.tdl")
        (lkb-pathname (this-directory) "mrsmunge.tdl"))
    (lkb-pathname (this-directory) "settings.lsp"))
(read-cached-leaf-types-if-available
    (list (lkb-pathname (parent-directory) "letypes.tdl")
        (lkb-pathname (parent-directory) "semrels.tdl")))
(read-cached-lex-if-available
    (lkb-pathname (parent-directory) "lexicon.tdl"))
(read-tdl-grammar-file-aux
    (lkb-pathname (parent-directory) "constructions.tdl"))
(read-morph-file-aux
    (lkb-pathname (this-directory) "inflr.tdl"))
(read-tdl-start-file-aux
    (lkb-pathname (parent-directory) "roots.tdl"))
(read-tdl-lex-rule-file-aux
```

```
(lkb-pathname (parent-directory) "lexrinst.tdl"))
(read-tdl-parse-node-file-aux
   (lkb-pathname (parent-directory) "parse-nodes.tdl"))
(lkb-load-lisp (this-directory) "mrs-initialization.lsp" t)
```

I won't go through this in detail, but note the following:

1. The command to read in the script file is specified to carry out all the necessary initializations of grammar parameters etc. So although it might look as though a script file can be read in via load like a standard Lisp file, this would cause various things to go wrong.

2. The first load statement looks for a file called Version.lsp in the directory above the one where the script file itself is located. (As before, all paths are given relative to the location of the script file, so the same script will work with different computers, provided the directory structure is maintained.) The file Version.lsp sets a variable that records the grammar version. This is used for record-keeping purposes and also to give names to the cache files (see §8.8).

3. The user preferences file (user-prefs.lsp) is loaded automatically. It is kept in the same directory as the other globals file, which allows a user to set up different preferences for different grammars.

4. The checkpaths.lsp file is loaded to improve efficiency, as discussed in §8.3.1. The third argument to lkb-load-lisp is t, to indicate that the file is optional.

5. The comlex.lsp file contains code which provides an interface to a lexicon constructed automatically from the COMLEX lexicon which is distributed by the Linguistic Data Consortium. This acts as a secondary lexicon when the specially built lexicon is missing a word entry.

6. There is a list of type files read in by read-tdl-type-files-aux — the second argument to this function is a file for display settings (see §6.1.9).

7. Two type files are specified as leaf types (see §8.7). The leaf types are cached so that they can be read in quickly if the files are unaltered.

8. The lexicon is cached so that it can be read in quickly if it is unaltered: see §8.8.

9. The final file mrs-initialization.lsp contains code to initialize the behaviour of the MRS code (if present). This is responsible for loading the grammar-specific MRS parameters file.

8.2.1 Loading functions

The following is a full list of available functions for loading in LKB source files written using the TDL syntax. All files are specified as full pathnames. Unless otherwise specified, details of file format etc are specified in Chapters 4 and 5 and error messages etc are in Chapter 7.

load-lkb-preferences *directory file-name*

Loads a preferences file and sets ***user-params-file*** to the name of that file, so that any preferences the user changes interactively will be reloaded next session.

read-tdl-type-files-aux *file-names* &optional *settings-file*

Reads in a list of type files and processes them. An optional settings file controls shrunkenness (see §6.1.9). If you wish to split types into more than one file, they must all be specified in the file name list, since processing assumes it has all the types (apart from leaf types).

read-tdl-leaf-type-file-aux *file-name*

Reads in a leaf type file. There may be more than one such command in a script. See §8.7.

read-cached-leaf-types-if-available *file-name(s)*

Takes a file or a list of files. Reads in a leaf type cache if available (WARNING, there is no guarantee that it will correspond to the file(s) specified). If there is no existing leaf type cache, or it is out of date, reads in the specified files using **read-tdl-leaf-type-file-aux**. See §8.8.

read-tdl-lex-file-aux *file-name*

Reads in a lexicon file. There may be more than one such command in a script.

read-cached-lex-if-available *file-name(s)*

Takes a file or a list of files. Reads in a cached lexicon if available (WARNING, there is no guarantee that it will correspond to the file(s) specified). If there is no existing cached lexicon, or it is out of date, reads in the specified files using **read-tdl-lex-file-aux**. See §8.8.

read-tdl-grammar-file-aux *file-name*

Reads in a grammar file. There may be more than one such command in a script.

read-morph-file-aux *file-name*

Reads in a lexical rule file with associated affixation information. Note that the morphology system assumes there will only be one such file in a grammar, so this command may not occur more than once in a script, even though it takes a single file as argument.

read-tdl-lex-rule-file-aux *file-name*

Reads in a lexical rule file where the rules do not have associated affixation information. There may be more than one such command in a script.

`load-irregular-spellings` *file-name*

Reads in a file of irregular forms. It is assumed there is only one such file in a grammar.

`read-tdl-parse-node-file-aux` *file-name*

Reads in a file containing entries which define parse nodes.

`read-tdl-start-file-aux` *file-name*

This command simply defines the entries, without giving them any particular functionality — this works because start symbols are enumerated in the globals file.

`read-tdl-psort-file-aux` *file-name*

This is actually defined in the same way as the previous command: it simply defines the entries, without giving them any particular functionality. It is retained for backward compatibility.

8.2.2 Utility functions

The following functions are defined as useful utilities for script files.

`this-directory`

Returns the directory which contains the script file (only usable inside the script file).

`parent-directory`

Returns the directory which is contains the directory containing the script file (only usable inside the script file).

`lkb-pathname` *directory name*

Takes a directory as specified by the commands above, and combines it with a file name to produce a valid full name for the other commands.

`lkb-load-lisp` *directory name* &optional *boolean*

Constructs a file name as with `lkb-pathname` and loads it as a Lisp file. If optional is t (i.e., true), ignore the file if it is missing, otherwise signals a (continuable) error.

8.3 Parsing and generation efficiency techniques

8.3.1 Check paths

The check paths mechanism greatly increases the efficiency of parsing and generation with large grammars like the LinGO ERG. It relies on the fact that unification failure during rule application is normally caused by type incompatibility on one of a relatively small set of paths. These paths can be checked very efficiently before full unification is attempted, thus providing a filtering mechanism which considerably improves performance. The paths are constructed automatically by batch parsing a representative range of sentences.

The menu command **Create quick check file**, described in 6.1.8, allows a set of check-paths to be created on the basis of a set of test sentences. To do the same thing non-interactively, the macro `with-check-path-list-collection` is used. It takes two arguments: the first is is a call to a batch parsing function. For example:

```
(with-check-path-list-collection "~aac/checkpaths.lsp"
  (parse-sentences "~aac/grammar/lkb/test-sentences"
    "~aac/grammar/lkb/results"))
```

The file of checkpaths created in this way is then read in as part of the script. For instance:

```
(lkb-load-lisp (this-directory) "checkpaths.lsp" t)
```

To maintain filtering efficiency, the checkpaths should be recomputed whenever there is a major change to the architecture of the TFSs used in the grammar. However, even if they are out-of-date, the checkpaths will never affect the result of parsing or generation, only the efficiency.

8.3.2 Key-driven parsing and generation

When the parser or generator attempts to apply a grammar-rule to two or more daughters, it is often the case that it is more efficient to check the daughters in a particular order, so that if unification is going to fail, it will do so as quickly as possible. This mechanism allows the grammar developer to specify a daughter as the *key*: the key daughter is checked first. The value of the path `*key-daughter-path*` in the daughter structure of the grammar rule should be set to the value of `*key-daughter-type*` when that daughter is the key. By default, the value of `*key-daughter-path*` is (KEY-ARG) and the value of `*key-daughter-type*` is +. So, for instance, if the daughters in the rule are described as a list which is the value of the feature ARGS and the first daughter is the key, the value of (ARGS FIRST KEY-ARG) should be +. A rule is not required to have a specified key and the `*key-daughter-path*` need not be present in this case. Specifying a key will never affect the result of parsing or generation.

8.3.3 Avoiding copying and tree reconstruction

The HPSG framework is usually described in such a way that phrases are complete trees: that is, a phrase TFS contains substructures which are also phrases. This leads to very large structures and computational inefficiency. However, HPSG also has a locality principle, which means that it is not possible for a phrase to access substructures that are daughters of its immediate daughters. Any such information has to be carried up explicitly.

The locality principle therefore guarantees that it is possible to remove the daughters of a rule after constructing the mother without affecting the result. The LKB system allows the grammar writer to supply a list of features which will not be passed from daughter to mother when parsing via the parameter *deleted-daughter-features*. In fact, in the samples grammars we have looked at in this book, *deleted-daughter-features* could have been set to (ARGS) to improve efficiency. This feature must be used with care, since if the grammar writer has not obeyed the locality principle with respect to the specified features, setting this parameter will cause different results to be obtained.[57]

8.3.4 Packing

Simple context free grammars can be parsed in cubic time because not all derivations need be explicitly computed. For instance, consider a sentence like:

(8.65)   The jack above the ace next to the queen is red.

The NP, *the jack above the ace next to the queen* could be analysed as having either of the following two structures:

(8.66)   the ((jack (above the ace)) (next to the queen))
(8.67)   the (jack (above (the (ace (next to the queen)))))

These have different semantics: in the first, the jack is above the ace and also next to the queen, while in the second, it is the ace that is next to the queen. However, if these structures were being accounted for by simple CFG rules, both would simply be an NP, and there could be no difference in how they subsequently interacted with the rest of the grammar. So a simple CFG parser can assume that there is only one edge for the NP, even though it can be derived in multiple different ways. In general, in a chart parser for simple CFGs, once an edge between two nodes n and m has been constructed, any subsequent edge with the same category is simply recorded but is not involved in further processing, because it would simply be duplicating the results from the first edge. This allows cubic time parsing, even for grammars where an exponential number of derivations is possible for a sentence.

With unification-based grammars, there are several potential problems in applying this technique. The first concerns the fact we are dealing with TFSs rather than simply atomic category symbols. This

---

[57]In many ways it would be more logical to define grammar rules with a separate mother feature, which is supported by the LKB system, but this doesn't fit in with the way that HPSG is generally described.

means that for maximal packing, the parser has to check for subsumption rather than equality. An edge which is more specific than an existing edge spanning the same nodes cannot result in any new parses, but further complexity arises because a new edge might be more general than an existing edge, in which case the new edge has to be checked but the existing edge could be frozen. This is discussed in detail by Oepen and Carroll (2000b).

The other problems concern the way the grammars are used. In the grammars we have seen, the structures that are produced for a node carry around information about their derivation, because of the ARGS. In terms of a CFG, it is as though the categories for the NPs above were not simply NP, but the following structures:

1. (NP (Det the) (N (N (N jack) (PP (P above) (NP (Det the) (N ace)))) (PP (P next-to) (NP (Det the) (N queen)))))
2. (NP (Det the) (N (N jack) (PP (P above) (NP (Det the) (N (N ace) (PP (P next-to) (NP (Det the) (N queen))))))))

Obviously these two structures are not the same. However, as we saw in the previous section, the locality principle of HPSG guarantees that the derivation information can be ignored because no required information is in the ARGS without also being coindexed with the rest of the phrase. Thus we can ignore information under ARGS when packing.

A related problem is that in many grammars, the semantics is built up in parallel with the syntax. This is convenient, but means that the two TFSs for 8.66 and 8.67 will be different. However, in the grammars used in this book and in the ERG, there is also a semantic locality principle, which guarantees that the internal structure of the list of elementary predications cannot be relevant to subsequent composition operations. This means that the semantics can be ignored too, for the purposes of packing, as long as we are prepared to reconstruct the TFSs when it is necessary to pass the semantic structures to another system component.

The LKB system contains code that implements packing, but it is not used by default. For further details and experimental results on the ERG, see Oepen and Carroll (2000b).

8.4   Irregular morphology

Irregular morphology may be specified in association with a particular lexical rule, in the way we saw in with g6morph in §5.2.1, but it is often more convenient to have a separate file for irregular morphemes. In the LKB, this file has to take the form of a Lisp string (i.e., it begins and ends with ")[58] containing irregular form entries, one per line, each consisting of a triplet:

1. inflected form
2. rule specification
3. stem

For example:

```
"
dreamt PAST-V_IRULE dream
fell PAST-V_IRULE fall
felt PAST-V_IRULE feel
gave PAST-V_IRULE give
"
```

The file is loaded with the command load-irregular-spellings in the script file. For instance:

```
(load-irregular-spellings
   (lkb-pathname (this-directory) "irregs.tab"))
```

The interpretation of irregular forms is similar to the operation of the regular morphology: that is, the irregular form is assumed to correspond to the spelling obtained when the specified rule is applied to a lexical entry with an orthography corresponding to the stem. The rule specification should be the identifier for a morphology rule.

The default operation of the irregular morphology system is one case where there may be an asymmetry in system behaviour between parsing and generation: when parsing, a regularly derived form will be accepted even if there is also an irregular form if the value of the LKB parameter *irregular-forms-only-p* (see §9.2.3) is nil. Thus, for example, *gived* will be accepted as well as *gave*, and *dreamed* as well as *dreamt*. If the value of *irregular-forms-only-p* is t, an irregular form will block acceptance of a regular form, so only *gave* would be accepted. Generation will always produce the irregular form if there is one: with the current version of the LKB it will simply produce the first form given in the irregulars file for a particular rule even if there are alternative spellings. For instance, assume the following is included in the irregulars file:

```
dreamed PAST-V_IRULE dream
dreamt PAST-V_IRULE dream
```

It is a good idea to include both forms for parsing if the parameter *irregular-forms-only-p* is true, because it allows both variants to

---

[58]This rather clunky format is used for compatibility with other systems.

be accepted, but for generation, only *dreamed* would be produced.

8.5 Multiword lexemes

The LKB system allows lexical items to be specified which have a value for their orthography which is a list of more than one string. These items are treated as multiword lexemes and the parser checks that all the strings are present before putting the lexical entry on the chart.

Multiword lexemes may have at most one affixation site. This is specified on a per-entry basis, via the user-definable function `find-infl-pos` (see §9.3). By default, this allows affixation on the rightmost element: e.g., *ice creams*. It can be defined in a more complex way, for instance to allow *attorneys general* or *kicked the bucket* to be treated as multiword lexemes.

This component of the LKB is under very active development at the time of writing, so please check the website for possible updates.

8.6 Parse ranking

The current LKB parser supports a simple mechanism for ordering parses and for returning the first parse only. The application of this mechanism is controlled by the variable `*first-only-p*` which may be set interactively, via the **Options / Set options** menu command. The weighting is controlled by two functions, `rule-priority` and `lex-priority`, which may be defined in the grammar-specific file `user-fns.lsp`, if desired. Since the mechanism is under active development, I don't propose to document it in detail here, but I suggest that anyone who wishes to experiment with this capability looks at the definitions of `rule-priority` and `lex-priority` in the `user-fns` file for the LinGO ERG. However, please check the LKB website for updates.

8.7 Leaf types

The notion of a leaf type is defined for efficiency in grammar loading. A leaf type is a type which is not required for the valid definition or expansion of any other type or type constraint. Specifically this means that a leaf type must meet the following criteria:

1. A leaf type has no daughters.
2. A leaf type may not introduce any new features on its constraint.
3. A leaf type may not be used in the constraint of another type.
4. A leaf type only has one 'real' parent type — it may have one or more template parents (see 9.2.1).

Under these conditions, much more efficient type checking is possible, and the type constraint description can be expanded on demand rather than being expanded when the type file is read in.

In the current version of the LinGO ERG, most of the terminal lexical types (i.e., those from which lexical entries inherit directly) are leaf types, as are most of the relation types. The latter class is particularly important, since it means that a lexicon can be of indefinite size and expanded on demand, while still using distinct types to represent semantic relations.

Various checks are performed on leaf types to ensure they meet the above criteria. However the tests cannot be completely comprehensive if the efficiency benefits of leaf types are to be maintained. If a type is treated as a leaf type when it does not fulfill the criteria above, unexpected results will occur. Thus the leaf type facility has the potential for causing problems which are difficult to diagnose and it should therefore be used with caution.

8.8   Caches

The cache functionality is provided to speed up reloading a grammar when leaf types or the lexicon have not changed.

Lexicon cache   Whenever a lexicon is read into the LKB, it is stored in an external file rather than in memory. Lexical entries are pulled in from this file as required. The caching facility saves this file and an associated index file so that they do not need to be recreated when the grammar is read in again if the lexicon has not changed. The location of these files is set by the `user-fns.lsp` function `set-temporary-lexicon-filenames`. The file names are stored in the parameters `*psorts-temp-file*` and `*psorts-temp-index-file*` (see §9.3.1). The default function uses the function `lkb-tmp-dir` to find a directory and names the files `templex` and `templex-index` respectively.   A more complex version of `set-temporary-lexicon-filenames` is given in the `user-fns.lsp` file in the LinGO ERG: this uses a parameter set by `Version.lsp` to name the files.

The script command `read-cached-lex-if-available` (see §8.2.1), takes a file or list of files as arguments. If a cached lexicon is available in the locations specified by `*psorts-temp-file*` and `*psorts-temp-index-file*`, and is more recent than the file(s) given as arguments to the function, then it will be used instead of reading in the specified files from scratch. The code does not check to see whether the cached lexicon was created from the specified files, and there are other ways in which the system can be fooled. Thus you should definitely not use this unless you have a sufficiently large lexicon that the reload time is annoying.

Leaf type cache A similar caching mechanism is provided for leaf types (§8.7). The parameter storing the name of the file is `*leaf-temp-file*` but like the lexicon cache files this is set by the function `set-temporary-lexicon-filenames`.

8.9 Using emacs with the LKB system

We recommend the use of emacs (either gnuemacs or XEmacs) with the LKB for editing the grammar files. Details of how to obtain emacs and set it up for use with the LKB are on the LKB website. The LKB menu command **Show source** requires that emacs be used as the editor. The emacs interface also puts some LKB menu commands on the top menu bar of the emacs window, for convenience, and a TDL editing mode is available. The LKB website also contains a brief guide to emacs commands for beginners.

8.10 YADU

The structures used in the LKB system are not actually ordinary TFSs, but typed default feature structures (TDFSs). So far in this book I have assumed non-default structures and monotonic unification, but this is actually just a special case of typed default unification. The full description of TDFSs and default unification is given in Lascarides and Copestake (1999) (henceforth L+C). The following notes are intended to supplement that paper.

Default information is introduced into the description language by /, followed by an indication of the persistence of the default. In terms of the current implementation, the supported distinctions in persistence are between defaults which are fully persistent and those which become non-default when an entry TFS is constructed (referred to as description persistence). The value of the description persistence indicator is given by the parameter `*description-persistence*` — the default is l (i.e., the lowercase letter l, standing for 'lexical', not the numeral 1).

The modification to the BNF given for the TDL syntax in §4.4.6 is as follows:

Conjunction → *DefTerm* | *DefTerm* **&** *Conjunction*
DefTerm → *Term* | *Term* /*identifier Term* | /*identifier Term*

It is not legal to have a default inside a default.

As an example, the following definition of **verb** makes the features PAST, PASTP and PASSP indefeasibly coindexed, and PASTP and PASSP defeasibly coindexed. The definition for **regverb** makes the value of PAST be ed, by default.

```
verb := *top* &
[ PAST *top* /l #pp,
  PASTP #p /l #pp,
  PASSP #p /l #pp ].
```

```
regverb := verb &
[ PAST /l "ed" ] .
```

Any entry using these types will be completely indefeasible, since the defaults have description persistence. Further examples of the use of defaults, following the examples in L+C, are given in the files in data/yadu_test, distributed with the LKB.

You should note that the description language specifies dual TFSs from which BasicTDFSs are constructed as defined in §3.5 of L+C. See also the discussion of the encoding of VALP in §4.2 of L+C.

If you view a feature structure which contains defaults, you will see three parts. The first is the indefeasible structure, the second the 'winning' defeasible structure, and the third the tail. (Unfortunately this is very verbose: a more concise representation will be implemented at some point.)

Notice that the current implementation assumes that defaults are only relevant with respect to an inheritance hierarchy: default constraints which are specified on types other than the root node of a structure are ignored when expanding the feature structure.

8.11 MRS

MRS (minimal recursion semantics) is a framework for computational semantics which is intended to simplify the design of algorithms for generation and semantic transfer, and to allow the representation of underspecified scope while being fully integrated with a typed feature structure representation. It is decribed in detail in Copestake et al (1999). The LKB system supports MRS in providing various procedures for processing MRS structures (for printing, checking correctness, scoping, translation to other formalisms etc). The LKB generation component (described in §5.4.4 and §8.12) assumes an MRS style of representation. The LinGO grammar produces MRS structures, but the example grammars discussed in this book produce a form of simplified MRS which is not adequate to support scope resolution, though it can be used for generation. The MRS menu commands are described in §6.1.5 and §6.4. Further documentation will become available via the LKB webpage.

8.12 Generation

The generator requires a grammar which is using MRS or a relatively similar formalism The generator is described in Carroll et al (1999). A few remarks may be helpful to supplement §5.4.4.

1. The grammars that use MRS contain a file defining some necessary global parameters. This is called `mrsglobals.lsp` for the example grammars and `mrsglobals-eng.lsp` for the LinGO ERG.

2. Before generating, the lexicon and lexical and grammar rules must be indexed. The function which does this, `index-for-generator`, can be run with the **Index** command from the **Generate** menu, or it can be directly run from the script file as in §5.4.4.

3. If the grammar uses lexical entries which do not have any relations, warning messages will be issued by the indexing process. A set of grammar-specific heuristics can be defined which will determine which such lexical entries might be required for a given semantics. These are loaded via the **Load Heuristics** command on the **Generate** menu. If no such heuristics are defined, the system attempts to use all null semantics items, which can be very slow.

4. The generator supports a special treatment of intersective modification, discussed in Carroll et al (1999). The LKB parameters which control this are listed in §9.2.5.

5. There is a spellout component to the generator, which currently simply looks after the *a/an* alternatives.

6. A LKB parameter, `*duplicate-lex-ids*`, is used to specify lexical entry identifiers which should be ignored, because they are simply alternative spellings of another lexical entry. This mechanism is expected to be refined in the near future.

8.13 Testing and Diagnosis

The batch parsing support which is distributed with the LKB is very simple. For extensive grammar development, much more sophisticated tools are available through the [incr tsdb()] package that can be used with the LKB. The following description is by Stephan Oepen (to whom all comments should be directed: see the website below for contact information).

The [incr tsdb()] package combines the following components and modules:

- test and reference data stored with annotations in a structured database; annotations can range from minimal information (unique test item identifier, item origin, length et al.) to fine-grained linguistic classifications (e.g., regarding grammaticality and linguistic phenomena presented in an item) as represented by the TSNLP test suites (Oepen et al, 1997).

- tools to browse the available data, identify suitable subsets and feed them through the analysis component of a processing system like the LKB or PAGE;

- the ability to gather a multitude of precise and fine-grained system performance measures (like the number of readings obtained per test item, various time and memory usage metrics, ambiguity and non-determinism parameters, and salient properties of the result structures) and store them as a *competence and performance profile* in the database;

- graphical facilities to inspect the resulting profiles, analyze system competence (i.e., grammatical coverage and overgeneration) and performance at highly variable granularities, aggregate, correlate, and visualize the data, and compare profiles obtained from previous grammar or system versions or other platforms.

- a universal pronounciation rule: *tee ess dee bee plus plus* is the name of the game.

Please see:

`http://www.coli.uni-sb.de/itsdb/`

for links to more information about the [incr tsdb()] package and details of how to get it.

8.14 Parse tree labels

The grammars in this book use a simple node labelling scheme in parse trees, as described in 4.5.7. However, if `*simple-tree-display*` is set to `nil` rather than `t`, the parse tree labelling is more complex, and this more complex scheme is outlined here.

There are two classes of templates: *label* and *meta* structures (see §9.1.3 for the parameters). Each label TFS specifies a single string at a fixed path in its TFS: LABEL-NAME in the ERG and textbook grammars. For instance, the following is a label from the textbook grammar.

```
np :=  label &
[ SYN [ HEAD noun,
        SPR < > ],
  LABEL-NAME "NP" ].
```

Meta templates are used for things such as / (e.g., to produce the label S/NP). If meta templates are specified, each meta template TFS must specify a prefix string and a suffix string (by default, at the paths META-PREFIX and META-SUFFIX). For instance, the following is specified as a meta template in the textbook grammar:

```
slash := meta &
  [ SYN [ GAP [ LIST ne-list ] ],
    META-PREFIX "/",
    META-SUFFIX "" ].
```

To calculate the label for a node, the label templates are first checked to find a match. Matching is tested by unification of the template feature structure, excluding the label path, with the TFS on the parse tree node. For instance, if a parse tree node had the following description:

```
[ SYN [ HEAD *top*,
        SPR < > ]].
```

it would unify with

```
[ SYN [ HEAD noun,
        SPR < > ]].
```

and could thus be labelled NP, given the structure above. There is a parameter, *label-fs-path*, which allows templates to be checked on only the substructure of the node TFS which follows that path, but this parameter is set to the empty path in the textbook grammar.

If meta templates are specified, the TFS at the end of the path specified by the parameter *recursive-path* ((SYN GAP LIST FIRST) in the textbook grammar) is checked to see whether it matches a meta template. If it does, the meta structure is checked against the label templates (the parameter *local-path* allows them to be checked against a substructure of the meta structure). The final node label is constructed as the concatenation of the first label name, the meta prefix, the label of the meta structure, and the meta suffix.

8.15   Linking the LKB to other systems

The LKB has a non-graphical user interface which can be used by people who for whatever reason can't run the graphical user interface. Apart from the graphical user interface, the LKB system mostly uses ANSI-standard Common Lisp. The tty interface is also useful when using the LKB in conjunction with another system. I won't list the tty commands here (though see 6.1.4 for do-parse-tty, which is the most generally used). Some information on the tty commands is given on the LKB website.

The easiest way to hook up the output of the LKB parser to another system is to use a semantic representation compatible with MRS. A range of different output options are available for MRS, and writing alternative output formats is reasonably straightforward. Code which links MRS output to a canonical form representation suitable for use with theorem provers is in the LKB source directory `tproving`.

A lower-level interface is available via user-defined functions which are specified as the value of the parameter `*do-something-with-parse*`. This allows access to the parser output in the form of TFSs, which can be manipulated via the interface functions defined in the LKB source code file `mrs/lkbmrs.lisp`.

Further details of interfacing to the LKB will be given on the website, as time permits!

9

Details of system parameters

Various parameters control the operation of the LKB and some feature and type names are regarded as special in various ways. The system provides default settings but typically, each grammar will have its own files which set these parameters, redefining some of the default settings, though a set of closely related grammars may share parameter files. Each script must load the grammar-specific parameter files before loading any of the other grammar files (see §4.5.1 and §8.2). Other parameters control aspects of the system which are not grammar specific, but which concern things such as font size, which are more a matter of an individual user's preferences. This class of parameters can mostly be set via the **Options** menu (described in §6.1.9). Changing parameters in the **Options** menu results in a file being automatically generated with the preferences: this shouldn't be manually edited, since any changes may be overridden. There's nothing to prevent global variables which affect user preferences being specified in the grammar-specific globals files, but it is better to avoid doing this, since it could be confusing.

There are also some functions which can be (re)defined by the grammar writer to control some aspects of system behavior.

This chapter describes all the parameters and functions that the LKB system allows the user/grammar writer to set, including both grammar specific and user preference parameters. We sometimes refer to the parameters as *globals*, since they correspond to Common Lisp global variables. Note that the parameters in the grammar-specific globals files are all specified as either:

(def-lkb-parameter global-variable value comment)

or

(defparameter global-variable value comment)

where the comment is optional. Thus only a very basic knowledge of

209

Common Lisp is required to edit a globals file: the main thing to remember is that symbols have to be preceded by a quote, for instance:

`(def-lkb-parameter *string-type* 'string)`

See §4.5.2 for a lightening introduction to Common Lisp syntax.

The descriptions below give the default values for the variables and functions (as they are set in the LKB source files `main/globals` and `main/user-fns`). The description of the global variables is divided into sections based on the function of the variables: these distinctions do not correspond to any difference in the implementation with the exception that the globals which can be set via the **Options/ Set options** menu are nearly all ones that are specified as being grammar independent. (Not all grammar independent variables are available under **Options**, since some are too complex to set interactively, or are rarely used.)

To set a value interactively, use the **Options/ Set options** command on the parameter.

9.1 Grammar independent global variables

9.1.1 System behaviour

gc-before-reload, nil — This boolean parameter controls whether or not a full garbage collection is triggered before a grammar is reloaded. It is best set to `t` for large grammars, to avoid image size increasing, but it is `nil` by default, since it slows down reloading.

sense-unif-fn, nil — If set, this must correspond to a function. See `make-sense-unifications` in §9.3, below.

maximum-number-of-edges, 500 — A limit on the number of edges that can be created in a chart, to avoid runaway grammars taking over multi-user machines. If this limit is exceeded, the following error message is generated:

`Error: Probable runaway rule: parse/generate aborted`
`(see documentation of *maximum-number-of-edges*)`

When parsing a short sentence with a small grammar, this message is likely to indicate a rule which is applying circularly (that is, applying to its own output in a way that will not terminate). But this value must be increased to at least 2000 for large scale grammars with a lot of ambiguity such as the LinGO ERG. May be set interactively.

maximum-number-of-tasks, 50000 — A limit on the number of tasks that can be put on the agenda. This limit is only likely to be exceeded because of an error such as a circularly applicable rule.

chart-limit, 100 — The limit on the number of words (actually tokens as constructed by the tokenizer function) in a sentence which can be parsed. Whether the system can actually parse sentences of this length depends on the grammar!

maximal-lex-rule-applications, 7 — The number of lexical rule applications which may be made before it is assumed that some rules are applying circularly and the system signals an error.

display-type-hierarchy-on-load, t — A boolean variable, which controls whether the type hierarchy is displayed on loading or not. This must be set to nil for grammars with large numbers of types, because the type hierarchy display becomes too slow (and if the type hierarchy involves much multiple inheritance, the results are not very readable anyway). May be changed interactively.

9.1.2 Display parameters

feature-ordering, nil — A list which is interpreted as a partial order of features for setting the display ordering when TFSs are displayed or printed. See §6.3.

show-morphology, t — A boolean variable. If set, the morphological derivations are shown in parse trees (see §6.5). May be set interactively.

show-lex-rules, t — A boolean variable. If set, applications of lexical rules are shown in parse trees (see §6.5). May be set interactively.

simple-tree-display, nil — A boolean variable which can be set in order to use the simple node labelling scheme in parse trees. See §4.5.7 and §8.14.

substantive-roots-p, nil — A boolean variable which can be set to allow the structures constructed when checking the start conditions to be regarded as real edges for the purposes of chart display.

display-type-hierarchy-on-load, t — see §9.1.1 above. May be set interactively.

parse-tree-font-size, 12 — The font size for parse tree nodes. May be set interactively.

fs-type-font-size, 12 — The font size for nodes in AVM windows. May be set interactively.

fs-title-font-size, 12 — The font size for titles in AVM windows. May be set interactively.

type-tree-font-size, 12 — The font size for the nodes in the type hierarchy. May be set interactively.

dialog-font-size, 12 — The font size used in dialogue windows. May be set interactively.

maximum-list-pane-items, 50 — The maximum number of rules, lexical entries etc that will be offered as choices in menus that allow selection of such entities.

9.1.3 Parse tree node labels

These parameters are only used when *simple-tree-display* is nil. For details of how this operates, see §8.14.

label-path, (LABEL-NAME) — The path where the name string is stored.

prefix-path, (META-PREFIX) — The path for the meta prefix symbol.

suffix-path, (META-SUFFIX) — The path for the meta suffix symbol.

recursive-path, (NON-LOCAL SLASH LIST FIRST) — The path for the recursive category.

local-path, (LOCAL) — The path inside the node to be unified with the recursive node.

label-fs-path, (SYNSEM) — The path inside the node to be unified with the label node.

label-template-type, label — The type for all label templates.

9.1.4 Defaults

See §8.10.

description-persistence, 1 — The symbol used to indicate that a default should be made hard (if possible) when an entry is expanded into a complete TFS.

9.2 Grammar specific parameters

9.2.1 Type hierarchy

toptype, top — This should be set to the value of the top type: *top* in the example grammars. See §3.2.

string-type, string — The name of the type which is special, in that all Lisp strings are recognised as valid subtypes of it. See §3.2.

9.2.2 Orthography and lists

orth-path, (orth lst) — The path into a sign, specified as a list of features, which leads to the orthographic specification. See §4.1 and also the functions make-sense-unifications and make-orth-tdfs in

§9.3.

list-head, (hd) — The path for the first element of a list. See §4.4.2.

list-tail, (tl) — The path for the rest of a list. See §4.4.2.

list-type, *list* — The type of a list. See §4.4.2.

empty-list-type, *null* — The type of an empty list — it must be a subtype of ***list-type***. See §4.4.2.

diff-list-type, *diff-list* — The type of a difference list (see §4.4.2).

diff-list-list, list — The feature for the list portion of a difference list. See §4.4.2.

diff-list-last, last — The feature for the last element of a difference list. See §4.4.2.

9.2.3 Morphology and parsing

lex-rule-suffix, nil — If set, this is appended to the string associated with an irregular form in the irregs file in order to construct the appropriate inflectional rule name. See §8.4. Required for PAGE compatibility in the LinGO ERG.

mother-feature, 0 — The feature specifying the mother in a rule. May be NIL (is nil with all grammars discussed in this book).

start-symbol, sign — A type which specifies the type of any valid parse. Can be used to allow relatively complex start conditions. See §4.5.6. Unlike most grammar specific parameters, this can be set interactively to allow a switch between parsing fragments and only allowing full sentences, for instance.

deleted-daughter-features, nil — A list of features which will not be passed from daughter to mother when parsing. This should be set if efficiency is a consideration in order to avoid copying parts of the TFS that can never be (directly) referenced from rules pertaining to higher nodes in the parse tree. This will include daughter features in HPSG, since it is never the case that a structure can be directly selected on the basis of its daughters. See §8.3.3.

key-daughter-path, (key-arg) — A path into a daughter in a rule which should have its value set to `*key-daughter-type*` if that daughter is to be treated as the key. See §8.3.2.

key-daughter-type, + — . See above.

check-paths, nil — An association list in which the keys are feature paths that often fail — these are checked first before attempting unification to improve efficiency. See §8.3.1. The value of this parameter could be set in the globals file, but since the list of paths is automatically generated, it is normally kept in a distinct file. The parameter should ideally be set to `nil` in the globals file for grammars which do not use check paths, in case the grammar is read in after a previous grammar which does set the paths.

check-path-count, 30 — The number of check paths which are actually used when parsing (see above): set empirically to give maximum performance.

irregular-forms-only-p, nil — If set, the parser will not invoke the morphological analyzer on a form which has an irregular spelling. This prevents the system analyzing words such as *mouses*. Note that this means that if the grammar writer wants to be able to analyze *dreamed* as well as *dreamt*, both entries have to be in the irregular morphology file. Also note that we currently assume (incorrectly) that spelling is not affected by sense. For instance, the system cannot handle the usage where the plural of *mouse* is *mouses* when referring to computers. Note that this flag does not affect generation, which always treats an irregular form as blocking a regular spelling. See §8.4.

first-only-p, nil — If set, only one parse will be returned, where any preferences must be defined as specified in §8.6. May be set interactively.

9.2.4   Compare parameters

discriminant-path, (synsem local cont key) — A path used by the Compare display to identify a useful discriminating position in a structure — see §6.1.4.

9.2.5   Generator parameters

gen-first-only-p, nil — If set only one realization will be returned, where any preferences must be defined as specified in §8.6. May be set interactively.

semantics-index-path, (synsem local cont index) — The path used by the generator to index the chart. See §5.4.4.

intersective-rule-names, (adjh_i nadj_i hadj_i_uns) — The names of rules that introduce intersective modifiers. Used by the generator to improve efficiency. Default value is appropriate for the LinGO grammar. It should be set to NIL for grammars where the intersective modifier rules do not meet the necessary conditions for adjunction. See §8.12 and also the function `intersective-modifier-dag-p` in §9.3.

duplicate-lex-ids, (an) — Used in grammars which do not have any way of representing alternative spellings, this is a list of lexical identifiers that should be ignored by the generator. (The *a/an* alternatives are chosen by a post-generation spelling realization step.) See §8.12.

9.2.6 MRS parameters

I will not go through most of the MRS parameters here because they are currently frequently being revised. Documentation will be available on the website. They are stored in a separate file from the other globals (the file is called `mrsglobals.lsp` for the grammars in this book and `mrsglobals-eng.lisp` for the LinGO grammar).

mrs::*scoping-call-limit*, 10000 — Controls the search space for scoping.

9.3 User definable functions

make-sense-unifications — If this function is defined, it takes three arguments so that the orthographic form of a lexical entry, its id and the language are recorded in an appropriate place in the TFS. The value of `*sense-unif-fn*` must be set to this function, if it is defined. The function is not defined by default.

The idea is that this function can be used to specify paths (such as `ORTH.LIST.FIRST`) which will have their values set to the orthographic form of a lexical entry. This allows the lexicon files to be more succinct. It is assumed to be exactly equivalent to specifying that the paths take particular atomic values in the lexical entry. For example, instead of writing:

```
teacher_1 := noun-lxm &
[ ORTH.LIST.FIRST "teacher",
  SEM.RELS.LIST.FIRST.PRED teacher1_rel] .
```

the function could be defined so it was only necessary to write:

```
teacher_1 := noun-lxm.
```

since the value of the orthography string and the semantic relation name are predictable from the identifying material.

make-orth-tdfs — A function used by the parser which takes a string and returns a TFS which corresponds to the orthographic part of a sign corresponding to that string. The default version assumes that the string may have spaces, and that the TFS will contain a list representation with each element corresponding to one word (i.e., sequence of characters separated by a space). For instance, the string `"ice cream"` would give rise to the structure

```
[ ORTH [ LST [ HD ice
              TL [ HD cream ]]]]
```

establish-linear-precedence — A function which takes a rule TFS and returns the top level features in a list structures so that the ordering corresponds to mother, daughter 1, daughter 2 ... daughter n. The default version assumes that the daughters of a rule are indicated by the features 1 through n. See §4.1.

spelling-change-rule-p — A function which takes a rule structure and checks whether the rule has an effect on spelling. It is used to prevent the parser trying to apply a rule which affects spelling and which ought therefore only be applied by the morphology. The current value of this function checks for the value of NEEDS-AFFIX being **true**. If this matches a rule, the parser will not attempt to apply this rule. (Note that the function will return false if the value of NEEDS-AFFIX is anything other than **true**, since the test is equality, not unifiability.) See §5.2.1.

redundancy-rule-p — Takes a rule as input and checks whether it is a redundancy rule, defined as one which is only used in descriptions and is not intended to be applied productively. See §5.2.1. For instance, the prefix *step-*, as in *stepfather*, has a regular meaning, but only applies to a small, fixed set of words. A redundancy rule for *step-* prefixation can be specified to capture the regularity and avoid redundancy, but it can only be used in the lexical descriptions of *stepfather* etc, and not applied productively. (There is nothing to prevent productive lexical rules being used in descriptions.)

The default value of this function checks for the value of PRODUCTIVE being **false**. If this matches a rule, the parser will not attempt to apply that rule. (Note that the function will return false if the value of PRODUCTIVE is anything other than **false**, since the test is equality, not unifiability.)

preprocess-sentence-string — The function takes an input string and preprocesses it for the parser. The result of this function should be a single string which is passed to a simple word identifier which splits the string into words (defined as things with a space between them). So minimally this function could be simply return its input string. However, by default some more complex processing is carried out here, in order to strip punctuation and separate *'s*. Thus, in effect, this function controls the tokenizer: see §4.1.

find-infl-poss — This function is called when a lexical item is read in, but only for lexical items which have more than one item in their orth value (i.e., multiword lexemes). It must be defined to take three arguments: a set of unifications (in the internal data structures), an orthography value (a list of strings) and a sense identifier. It should return an integer, indicating which element in a multi-word lexeme may be inflected (counting left to right, leftmost is 1) or `nil`, which indicates that no item may be inflected. The default value for the function allows inflection on the rightmost element. See §8.5.

hide-in-type-hierarchy-p — Can be defined so that certain types are not shown when a hierarchy is displayed (useful for hierarchies where there are a very large number of similar leaf types, for instance representing semantic relations).

rule-priority — See §8.6. The default function assigns a priority of 1 to all rules.

lex-priority — See §8.6. The default function assigns a priority of 1 to all lexical items.

intersective-modifier-dag-p — Used by the generator to test whether a structure is an intersective modifier. Default value is applicable for the LinGO grammar. It should be set to NIL in grammars where intersective modifiers do not meet the conditions the generator requires for adjunction. See also the parameter `*intersective-rule-names*` in §9.2.5.

9.3.1 System files

There are two user definable functions which control two system files. The file names are associated with two global variables — these are initially set to `nil` and are then instantiated by the functions. The global variables are described below, but should not be changed by the user. The functions which instantiate them may need to be changed for different systems.

lkb-tmp-dir — This function attempts to find a sensible directory for the temporary files needed by the LKB. The default value for this on Unix is a directory `tmp` in the user's home directory: on a Macintosh it is `Macintosh HD:tmp`. The function should be redefined as necessary to give a valid path. It is currently only called by the function `set-temporary-lexicon-filenames` (below).

set-temporary-lexicon-filenames — This function is called in order to set the temporary files. It uses lkb-tmp-dir, as defined above. It is useful to change the file names in this function if one is working with multiple grammars and using caching to ensure that the lexicon file associated with a grammar has a unique name which avoids overriding another lexicon (see §8.8).

The files are defined by the following variables:

psorts-temp-file This file is constructed by the system and used to store the unexpanded lexical entries, in order to save memory. Once a lexical entry is used, it will be cached until either a new lexicon is read in, or until the **Tidy up** command is used (§6.1.8). If the temporary lexicon file is deleted or modified while the LKB is running, it will not be possible to correctly access lexical entries. The file is retained after the LKB is exited so that it may be reused if the lexicon has not been modified (see §8.8, §8.2.1 and the description of *psorts-temp-index-file*, below).

The pathname is actually specified as:

```
(make-pathname :name "templex"
               :directory (lkb-tmp-dir))
```

psorts-temp-index-file This file is used to store an index for the temporary lexicon file. If the option is taken to read in a cached lexicon (see §8.8 and §8.2.1), then the lexicon index is reconstructed from this file. If this file has been deleted, or is apparently outdated, the lexicon will be reconstructed from the source file.

leaf-temp-file This file is used to store cached leaf types.

References

Aït-Kaci, Hassan. (1984). A lattice-theoretic approach to computation based on a calculus of partially ordered type structures. University of Pennsylvania, Doctoral Dissertation.

Allwood, Jens, Lars-Gunnar Andersson and Östen Dahl. (1977). *Logic in Linguistics.* Cambridge University Press.

Alshawi, Hiyan (editor). (1992). *The Core Language Engine.* MIT Press, Cambridge, MA.

Bresnan, Joan and Ronald M. Kaplan. (1982). "Lexical-Functional Grammar: a formal system for grammatical representation." In *The Mental Representation of Grammatical Relations,* edited by Joan Bresnan. MIT Press.

Briscoe, Ted, Claire Grover, Bran Boguraev and John Carroll. (1987). A formalism and environment for the development of a large grammar of English. *Proceedings of the 10th International Joint Conference on Artificial Intelligence (IJCAI-87),* Milan, Italy: 703–708.

Briscoe, Ted, Ann Copestake and Valeria de Paiva. (1993). *Inheritance, defaults and the lexicon.* Cambridge University Press.

Butt, Miriam, Tracy Holloway King, María-Eugenia Niño and Frédérique Segond. (1999). *A Grammar Writer's Cookbook.* CSLI Publications, Stanford.

Calder, Jo. (1987). "Typed unification for natural language processing." In *Categories, polymorphism and unification,* edited by Ewan Klein and Johan van Benthem. Centre for Cognitive Science, University of Edinburgh, 65–72.

Callmeier, Ulrich. (2000). PET — A Platform for Experimentation with Efficient HPSG Processing Techniques. *Journal of Natural Language Engineering. Special Issue on Efficient Processing with HPSG: Methods, Systems, Evaluation* 6(1).

Carpenter, Bob. (1992). *The logic of typed feature structures*. (Tracts in Theoretical Computer Science), Cambridge University Press, Cambridge, England.

Carroll, John, Ann Copestake, Daniel Flickinger and Victor Poznanski. (1999). An Efficient Chart Generator for (Semi-)Lexicalist Grammars. *Proceedings of the 7th European Workshop on Natural Language Generation (EWNLG'99)*, Toulouse: 86–95.

Colmerauer, Alain. (1970). "Les systèmes-Q ou un formalisme pour analyser et synthétiser des phrases sur ordinateur." Publication Interne 43, Département d'Informatique, Université de Montréal, Canada.

Copestake, Ann. (1992). The ACQUILEX LKB: representation issues in semi-automatic acquisition of large lexicons. *Proceedings of the 3rd Conference on Applied Natural Language Processing (ANLP-92)*, Trento, Italy: 88–96.

Copestake, Ann. (1993). "The Compleat LKB." Technical report 316, University of Cambridge Computer Laboratory.

Copestake, Ann, Daniel Flickinger, Ivan A. Sag and Carl Pollard. (1999). "Minimal Recursion Semantics: An introduction." ms. CSLI, Stanford: http://www-csli.stanford.edu/~aac/papers/mrs.pdf.

Copestake, Ann and Daniel Flickinger. (2000). An open-source grammar development environment and broad-coverage English grammar using HPSG. *Proceedings of the Second conference on Language Resources and Evaluation (LREC-2000)*, Athens, Greece: 591–600.

Emele, Martin. (1994). The Typed Feature Structure Representation Formalism. *Proceedings of the International Workshop on Sharable Natural Language Resources*, Ikoma, Nara, Japan.

Emele, Martin and Remi Zajac. (1990). Typed unification grammars. *Proceedings of the 13th International Conference on Computaional Linguistics (COLING-90)*, Helsinki: 293–298.

Flickinger, Daniel. (1987). "Lexical rules in the hierarchical lexicon." PhD thesis, Stanford University.

Flickinger, Daniel, Carl Pollard and Thomas Wasow. (1985). Structure sharing in lexical representation. *Proceedings of the 23rd Annual Meeting of the Association for Computational Linguistics (ACL-85)*, University of Chicago: 262–268.

Flickinger, Daniel, Stephan Oepen, Hans Uszkoreit and Jun'ichi Tsujii (editors). (2000). Special Issue on Efficient Processing with HPSG: Methods, Systems, Evaluation. *Journal of Natural Language Engineering* 6(1).

Gazdar, Gerald. (1981). Unbounded dependencies and coordinate structure. *Linguistic Inquiry* 12: 155–184.

Gazdar, Gerald, Ewan Klein, Geoffrey Pullum and Ivan Sag. (1985). *Generalized Phrase Structure Grammar.* Basil Blackwell, Oxford.

Jurafsky, Daniel and James H. Martin. (2000). *Speech and language processing.* Prentice Hall.

Karttunen, Lauri. (1986). D-PATR: a development environment for unification-based grammars. *Proceedings of the 11th International Conference on Computational Linguistics (COLING-86)*, Bonn, Germany: 74–80.

Kasper, Robert T. and William C. Rounds. (1986). A logical semantics for feature structures. *Proceedings of the 24th Annual Conference of the Association for Computational Linguistics (ACL-86)*, Columbia University: 235–242.

Kasper, Robert T. and William C. Rounds. (1990). The logic of unification in grammar. *Linguistics and Philosophy* 13(1): 35–58.

Kay, Martin. (1979). Functional Grammar. *Proceedings of the Fifth Annual Meeting of the Berkeley Linguistic Society*, 142–158: Berkeley, CA, USA.

Kiefer, Bernd, Hans-Ulrich Krieger, John Carroll and Robert Malouf. (1999). A bag of useful techniques for efficient and robust parsing. *Proceedings of the 37th Annual Meeting of the Association for Computational Linguistics (ACL-99)*, 473–480: University of Maryland, USA.

King, Paul J.. (1989). A logical formalism for Head-Driven Phrase Structure Grammar. PhD dissertation, University of Manchester.

King, Paul J.. (1994). "An expanded logical formalism for Head-Driven Phrase Structure Grammar." Arbeitspapiere des SFB 340 59, Universitat Tubingen.

Krieger, Hans-Ulrich and Ulrich Schafer. (1994). TDL — A type description language for constraint-based grammars. *Proceedings of the 15th International Conference on Computational Linguistics (COLING-94)*, Kyoto, Japan: 893–899.

Lascarides, Alex and Ann Copestake. (1999). Default representation in constraint-based frameworks. *Computational Linguistics* 25 (1): 55–106.

Malouf, Robert, John Carroll and Ann Copestake. (2000). Efficient feature structure operations without compilation. *Journal of Natural Language Engineering (Special Issue on Efficient processing with HPSG: methods, systems, evaluation)* 6(1).

Moens, Marc, Jo Calder, Ewan Klein, Mike Reape, and Henk Zeevat. (1989). Expressing generalizations in unification-based grammar formalisms. *Proceedings of the 4th Conference of the European Chap-*

ter of the Association for Computational Linguistics (EACL-89), Manchester, England: 66–71.

Oepen, Stephan, Klaus Netter and Judith Klein. (1997). "TSNLP — Test Suites for Natural Language Processing." In *Linguistic Databases,* edited by John Nerbonne. CSLI Lecture Notes 77, CSLI Publications, Stanford CA, 13 – 36.

Oepen, Stephan and Daniel Flickinger. (1998). Towards Systematic Grammar Profiling. Test Suite Technology Ten Years After. *Journal of Computer Speech and Language: Special Issue on Evaluation* 12 (4): 411-437.

Oepen, Stephan and John Carroll. (2000a). Parser engineering and performance profiling. *Journal of Natural Language Engineering (Special Issue on Efficient processing with HPSG: methods, systems, evaluation)* 6(1).

Oepen, Stephan and John Carroll. (2000b). Ambiguity Packing in Constraint-based Parsing. Practical Results. *Proceedings of the First Conference of the North American Chapter of the ACL,* Seattle, WA, USA.

Oepen, Stephan, Klaus Netter and Hans Uzskoreit. (1997). "PAGE: A Platform for Advanced Grammar Engineering."
http://www.dfki.de/lt/systems/page/.

de Paiva, Valeria. (1993). "Types and Constraints in the LKB." In *Inheritance, Defaults and the Lexicon,* edited by E. J. Briscoe, A. Copestake and V. de Paiva. Cambridge University Press, Cambridge, England, 164–189.

Pereira, Fernando and David H.D. Warren. (1980). Definite Clause Grammars for Language Analysis —A Survey of the Formalism and a Comparison with Augmented Transition Networks. *Artificial Intelligence* 13:3: 231–278.

Pollard, Carl and Ivan A. Sag. (1987). *An information-based approach to syntax and semantics: Volume 1 fundamentals.* CSLI Lecture Notes 13, CSLI Publications, Stanford CA.

Pollard, Carl and Ivan A. Sag. (1994). *Head-driven phrase structure grammar.* Chicago University Press, Chicago.

Pulman, Stephen G.. (1996). Unification encodings of grammatical notations. *Computational Linguistics* 22: 295 – 328.

Pustejovsky, James. (1995). *The Generative Lexicon.* MIT Press, Cambridge, Mass.

Sag, Ivan A. and Thomas Wasow. (1999). *Syntactic Theory — a formal introduction.* CSLI Publications, Stanford, CA, USA.

Sanfilippo, Antonio. (1993). "LKB encoding of lexical knowledge." In *Inheritance, defaults and the lexicon,* edited by E. J. Briscoe,

A. Copestake and V. de Paiva. Cambridge University Press, Cambridge, England, 190–222.

Shieber, Stuart. (1986). *An introduction to unification-based approaches to grammar.* CSLI Publications (CSLI Lecture Notes No. 4), Stanford, CA, USA.

Shieber, Stuart Hans Uszkoreit, Fernando Pereira, Jane Robinson, and Mabry Tyson. (1983). "The formalism and implementation of PATR-II." In *Research on Interactive Acquisition and Use of Knowledge,* edited by Barbara J. Grosz and Mark E. Stickel. AI Center, SRI International, Menlo Park, CA, USA, 39–79.

Smolka, Gert. (1989). "Feature constraint logic for unification grammars." IWBS Report 93, IWBS-IBM, Stuttgart, Germany.

Tallerman, Maggie. (1998). *Understanding Syntax.* Arnold, London.

Uszkoreit, Hans, Rolf Backofen, Stephan Busemann, Abdel Kader Diagne, Elizabeth A. Hinkelman, Walter Kasper, Bernd Kiefer, Hans-Ulrich Krieger, Klaus Netter, Günter Neumann, Stephan Oepen and Stephen P. Spackman. (1994). DISCO — An HPSG-based NLP System and its Application for Appointment Scheduling. *Proceedings of the 15th International Conference on Computational Linguistics (COLING-94),* Kyoto, Japan.

Wood, Mary M.. (1993). *Categorial Grammars.* Routledge, London and New York.

Zeevat, Henk, Ewan Klein and Jo Calder. (1987). "An introduction to unification categorial grammar." In *Categorial grammar, unification grammar, and parsing: working papers in cognitive science, Volume 1,* edited by Nick Haddock, Ewan Klein and Glyn Morrill. Centre for Cognitive Science, University of Edinburgh, 195–222.

Index

Index of Menu Commands

Index of Parameters and Functions

CD REQUEST FORM

If you are unable to download the LKB system from the website, a Windows version can be provided on CD-ROM. To obtain this, please send a self-addressed, stamped CD mailer to:

CSLI Publications
attn: LKB System Disk
Ventura Hall
Stanford University
Stanford, CA 94305-4115
USA

This offer is limited to one CD per copy of the book: please enclose this page with your request.

Note the system requirements described in Chapter 2 of this book, and that this offer is made under the same licence conditions that apply to the LKB system generally: in particular, there is no warranty of any kind.